ANALYTICAL ELECTROCHEMISTRY

ANALYTICAL ELECTROCHEMISTRY

Third Edition

Joseph Wang

A JOHN WILEY & SONS, INC., PUBLICATION

Published by John Wiley & Sons, Inc., Hoboken, New Jersey.
Published simultaneously in Canada.

For general information on our other products and services or for technical support, please contact our Customer Care Department within the United States at (800) 762-2974, outside the United States at(317) 572-3993 or fax (317) 572-4002.

Wiley also publishes its books in a variety of electronic formats. Some content that appears in print may not be available in electronic formats. For more information about Wiley products, visit our web site at www.wiley.com.

Library of Congress Cataloging-in-Publication Data:
Wang, Joseph, 1948–
Analytical electrochemistry / Joseph Wang.—3rd ed.
p. cm.
ISBN-13 978-0-471-67879-3 (cloth)
ISBN-10 0-471-67879-1 (cloth)
1. Electrochemical analysis. I. Title.
QD115.W33 2006
543′.4—dc22

2005028185

Printed in the United States of America.

10 9 8 7 6

Dedicated to the memory of my parents, Elka and Moshe Wang

CONTENTS

PREFACE

The goal of this textbook is to cover the full scope of modern electroanalytical techniques and devices. The main emphasis is on electroanalysis, rather than physical electrochemistry. The objective is to provide a sound understanding of the fundamentals of electrode reactions and the principles of electrochemical methods, and to demonstrate their potential for solving real-life analytical problems. The high performance, small size, and low cost of electrochemical devices has led to many important detection systems. Given the impressive progress in electroanalytical chemistry and its growing impact on analytical chemistry, this work offers also an up-to-date, easy-to-read presentation of more recent advances, including new methodologies, sensors, detectors, and microsystems. The book is suitable for a graduate-level course in electroanalytical chemistry or as a supplement to a high-level undergraduate course in instrumental analysis. It should also be very useful to those considering the use of electroanalysis in their laboratories.

The material is presented in six roughly equal chapters. The first chapter is devoted to fundamental aspects of electrode reactions and the structure of the interfacial region. Chapter 2 discusses the study of electrode reactions and high-resolution surface characterization. Chapter 3 gives an overview of finite-current-controlled potential techniques. Chapter 4 describes the electrochemical instrumentation and electrode materials (including new and modified microelectrodes). Chapter 5 deals with the principles of potentiometric measurements and various classes of ion-selective electrodes, while Chapter 6 is devoted to the growing field of chemical sensors (including modern biosensors, gas sensors, microchip devices, and sensor arrays). Numerous up-to-date references, covering the latest literature, are given at the end of each chapter.

By discussing more recent advances, this book attempts to bridge the common gap between research literature and standard textbooks.

This third edition of *Analytical Electrochemistry* is extensively revised and updated, and reflects the rapid growth of electroanalytical chemistry since 1999. It contains a number of new topics, including DNA biosensors, impedance spectroscopy, detection for capillary electrophoresis, diamond electrodes, carbon-nanotube- and nanoparticle-based assays and devices, large-amplitude AC voltammetry, microfluidic ("lab on a chip") devices, or molecularly-imprinted polymeric sensors. Other topics, such as the principles of potentiometric measurements, spectroelectrochemistry, electrochemiluminescence, modified and microelectrodes, scanning electrochemical and atomic force microscopies, electrical communication between redox enzymes and electrodes, explosive detection, or enzyme and immunoelectrodes, have been greatly expanded. The entire text has been updated to cover the very latest (as of August 2005) developments in electroanalytical chemistry. Numerous new illustrations, worked-out examples and end-of-chapter problems have been added to this edition. Existing figures have been redrawn and improved. In the 5 years since the second edition I have received numerous suggestions, many of which have been incorporated in the second edition.

Finally, I wish to thank my wife, Ruth, and my daughter, Sharon, for their love and patience; Vairavan Subramanian and Daphne Hui for their technical assistance; the editorial and production staff of John Wiley & Sons, Inc. for their help and support; Professor Erno Pretsch (ETH, Zurich) for extremely useful suggestions; and the numerous electrochemists across the globe who led to the advances reported in this textbook. Thank you all!

JOSEPH WANG

Tempe, AZ

ABBREVIATIONS AND SYMBOLS

a	Activity
A	Area of electrode
Ab	Antibody
AC	Alternating current
AdSV	Adsorptive stripping voltammetry
AE	Auxiliary electrode
AES	Auger electron spectroscopy
AFM	Atomic force microscopy
Ag	Antigen
ASV	Anodic stripping voltammetry
B	Adsorption coefficient
BDD	Boron-doped diamond
C	Concentration
C_{dl}	Differential capacitance
CE	Counter electrode
CME	Chemically modified electrode
CNT	Carbon nanotube
CSV	Cathodic stripping voltammetry
CV	Cyclic voltammetry
CWE	Coated-wire electrode
CZE	Capillary-zone electrophoresis
D	Diffusion coefficient
DME	Dropping mercury electrode
DNA	Deoxyribonucleic acid
DP	Differential pulse

DPV	Differential pulse voltammetry
E	Potential (V)
ΔE	Pulse amplitude
E°	Standard electrode potential
$E_{1/2}$	Half-wave potential
E_{p}	Peak potential
E_{pzc}	Potential of zero charge
EC	Electrode process involving electrochemical followed by chemical steps
ECL	Electrochemiluminescence
EQCM	Electrochemical quartz crystal microbalance
ESCA	Electron spectroscopy for chemical analysis
EXAFS	X-ray adsorption fine structure
F	Faraday constant
FET	Field effect transistor
FIA	Flow injection analysis
f_i	Activity coefficient
f_0	Base resonant frequency
FTIR	Fourier transform infrared
$\Delta G‡$	Free energy of activation
HMDE	Hanging mercury drop electrode
i	Electric current
i_{c}	Charging current
i_{t}	Tunneling current
IHP	Inner Helmholz plane
IRS	Internal reflectance spectroscopy
ISE	Ion-selective electrode
ISFET	Ion-selective field effect transistor
J	Flux
k_{ij}^{pot}	Potentiometric selectivity coefficient
k°	Standard rate constant
K_{m}	(1) Michaelis Menten constant; (2) mass transport coefficient
LB	Langmuir–Blodgett
LCEC	Liquid chromatography/electrochemistry
LEED	Low-energy electron diffraction
m	Mercury flow rate (in polarography)
Δm	Mass charge (in EQCM)
MFE	Mercury film electrode
μTAS	Micro–total analytical system
MIP	Molecularly imprinted polymer
MLR	Multiple linear regression
MWCNT	Multiwall carbon nanotube
N	Collection efficiency
n	Number of electrons transferred
NP	Normal pulse

O	The oxidized species
OHP	Outer Helmholz plane
OTE	Optically transparent electrode
PAD	Pulsed amperometric detection
PCR	Principal-component regression
PLS	Partial least squares
PSA	Potentiometric stripping analysis
PVC	Poly(vinyl chloride)
q	Charge
QCM	Quartz crystal microbalance
R	(1) Resistance; (2) gas constant
R_p	Electron transfer resistance
R_s	Ohmic resistance of the electrolyte solution
RDE	Rotating disk electrode
Re	Reynolds number
RE	Reference electrode
RRDE	Rotating ring–disk electrode
RVC	Reticulated vitreous carbon
S	(1) Barrier width (in STM); (2) substrate
SAM	Self-assembled monolayer
SECM	Scanning electrochemical microscopy
SEM	Scanning electron microscopy
SERS	Surface enhanced Raman scattering
SPM	Scanning probe microscopy
STM	Scanning tunneling microscopy
SW	Square wave
SWCNT	Single-wall carbon nanotube
SWV	Square-wave voltammetry
T	Temperature
t	Time
t_m	Transition time (in PSA)
U	Flow rate
UHV	Ultrahigh vacuum
v	Potential scan rate
V_{Hg}	Volume of mercury electrode
W□	Peak width (at half-height)
WE	Working electrode
WJD	Wall-jet detector
XPS	X-ray photoelectron spectroscopy
α	Transfer coefficient
Γ	Surface coverage
γ	Surface tension
δ	Thickness of the diffusion layer
δ_H	Thickness of the hydrodynamic boundary layer

ε	Dielectric constant
η	Overvoltage
μ	Ionic strength
ν	Kinematic viscosity
ω	Angular velocity

1

FUNDAMENTAL CONCEPTS

1.1 WHY ELECTROANALYSIS?

Electroanalytical techniques are concerned with the interplay between electricity and chemistry, namely, the measurements of electrical quantities, such as current, potential, or charge and their relationship to chemical parameters. Such use of electrical measurements for analytical purposes has found a vast range of applications, including environmental monitoring, industrial quality control, or biomedical analysis. Advances since the mid-1980s, including the development of ultramicroelectrodes, the design of tailored interfaces and molecular monolayers, the coupling of biological components and electrochemical transducers, the synthesis of ionophores and receptors containing cavities of molecular size, the development of ultratrace voltammetric techniques or of high-resolution scanning probe microscopies, and the microfabrication of molecular devices or efficient flow detectors, have led to a substantial increase in the popularity of electroanalysis and to its expansion into new phases and environments. Indeed, electrochemical probes are receiving a major share of the attention in the development of chemical sensors.

In contrast to many chemical measurements, which involve homogeneous bulk solutions, electrochemical processes take place at the electrode–solution interface. The distinction between various electroanalytical techniques reflects the type of electrical signal used for the quantitation. The two principal types

Analytical Electrochemistry, Third Edition, by Joseph Wang

of electroanalytical measurements are potentiometric and potentiostatic. Both types require at least two electrodes (conductors) and a contacting sample (electrolyte) solution, which constitute the electrochemical cell. The electrode surface is thus a junction between an ionic conductor and an electronic conductor. One of the two electrodes responds to the target analyte(s) and is thus termed the *indicator* (or *working*) electrode. The second one, termed the *reference* electrode, is of constant potential (i.e., independent of the properties of the solution). Electrochemical cells can be classified as *electrolytic* (when they consume electricity from an external source) or *galvanic* (if they are used to produce electrical energy).

Potentiometry (discussed in Chapter 5), which is of great practical importance, is a static (zero-current) technique in which the information about the sample composition is obtained from measurement of the potential established across a membrane. Different types of membrane materials, possessing different ion recognition processes, have been developed to impart high selectivity. The resulting potentiometric probes have thus been widely used for several decades for direct monitoring of ionic species such as protons or calcium, fluoride, and potassium ions in complex samples.

Controlled-potential (potentiostatic) techniques deal with the study of charge transfer processes at the electrode–solution interface, and are based on dynamic (non-zero-current) situations. Here, the electrode potential is being used to derive an electron transfer reaction and the resultant current is measured. The role of the potential is analogous to that of the wavelength in optical measurements. Such a controllable parameter can be viewed as "electron pressure," which forces the chemical species to gain or lose an electron (reduction or oxidation, respectively). Accordingly, the resulting current reflects the rate at which electrons move across the electrode–solution interface. Potentiostatic techniques can thus measure any chemical species that is electroactive, that is, that can be made to reduce or oxidize. Knowledge of the reactivity of functional group in a given compound can be used to predict its electroactivity. Nonelectroactive compounds may also be detected in connection with indirect or derivatization procedures.

The advantages of controlled-potential techniques include high sensitivity, selectivity toward electroactive species, a wide linear range, portable and low-cost instrumentation, speciation capability, and a wide range of electrodes that allow assays of unusual environments. Several properties of these techniques are summarized in Table 1.1. Extremely low (nanomolar) detection limits can be achieved with very small (5–20-μL) sample volumes, thus allowing the determination of analyte amounts ranging from 10^{-13} to 10^{-15} mol on a routine basis. Improved selectivity may be achieved via the coupling of controlled-potential schemes with chromatographic or optical procedures.

This chapter attempts to give an overview of electrode processes, together with discussion of electron transfer kinetics, mass transport, and the electrode–solution interface.

TABLE 1.1 Properties of Controlled-Potential Techniques[a]

Technique	Working Electrode	Detection Limit (M)	Speed (Time per Cycle) (min)	Response Shape
DC polarography	DME	10^{-5}	3	Wave
NP polarography	DME	5×10^{-7}	3	Wave
DP polarography	DME	10^{-8}	3	Peak
DP voltammetry	Solid	5×10^{-7}	3	Peak
SW polarography	DME	10^{-8}	0.1	Peak
AC polarography	DME	5×10^{-7}	1	Peak
Chronoamperometry	Stationary	10^{-5}	0.1	Transient
Cyclic voltammetry	Stationary	10^{-5}	0.1–2	Peak
Stripping voltammetry	HMDE, MFE	10^{-10}	3–6	Peak
Adsorptive stripping voltammetry	HMDE	10^{-10}	2–5	Peak
Adsorptive stripping voltammetry	Solid	10^{-9}	4–5	Peak
Adsorptive catalytic stripping voltammetry	HMDE	10^{-12}	2–5	Peak

[a] All acronyms used here are included in the "Abbreviations and Symbols" list following the Preface.

1.2 FARADAIC PROCESSES

The objective of controlled-potential electroanalytical experiments is to obtain a current response that is related to the concentration of the target analyte. Such an objective is accomplished by monitoring the transfer of electron(s) during the redox process of the analyte:

$$\mathrm{O} + ne^- \rightleftharpoons \mathrm{R} \tag{1.1}$$

where O and R are the oxidized and reduced forms, respectively, of the redox couple. Such a reaction will occur in a potential region that makes the electron transfer thermodynamically or kinetically favorable. For systems controlled by the laws of thermodynamics, the potential of the electrode can be used to establish the concentration of the electroactive species at the surface [$C_{\mathrm{O}}(0,t)$ and $C_{\mathrm{R}}(0,t)$] according to the Nernst equation

$$E = E^\circ + \frac{2.3RT}{nF}\log\frac{C_{\mathrm{O}}(0,t)}{C_{\mathrm{R}}(0,t)} \tag{1.2}$$

where E° is the standard potential for the redox reaction, R is the universal gas constant (8.314 J K^{-1} mol^{-1}), T is the Kelvin temperature, n is the number of electrons transferred in the reaction, and F is the Faraday constant [96,487 C (coulombs)]. On the negative side of E°, the oxidized form thus tends to be reduced, and the forward reaction (i.e., reduction) is more favorable. The current resulting from a change in oxidation state of the electroactive species is termed the *faradaic current* because it obeys Faraday's law (i.e., the reaction of 1 mol of substance involves a change of $n \times 96{,}487$ C). The faradaic current is a direct measure of the rate of the redox reaction. The resulting current–potential plot, known as the *voltammogram*, is a display of current signal [vertical axis (ordinate)] versus the excitation potential [horizontal axis (abscissa)]. The exact shape and magnitude of the voltammetric response is governed by the processes involved in the electrode reaction. The total current is the summation of the faradaic currents for the sample and blank solutions, as well as the nonfaradaic charging background current (discussed in Section 1.3).

The pathway of the electrode reaction can be quite complicated, and takes place in a sequence that involves several steps. The rate of such reactions is determined by the slowest step in the sequence. Simple reactions involve only mass transport of the electroactive species to the electrode surface, electron transfer across the interface, and transport of the product back to the bulk solution. More complex reactions include additional chemical and surface processes that either precede or follow the actual electron transfer. The net rate of the reaction, and hence the measured current, may be limited by either mass transport of the reactant or the rate of electron transfer. The more sluggish process will be the rate-determining step. Whether a given reaction is controlled by mass transport or electron transfer is usually determined by the type of compound being measured and by various experimental conditions (electrode material, media, operating potential, mode of mass transport, time scale, etc.). For a given system, the rate-determining step may thus depend on the potential range under investigation. When the overall reaction is controlled solely by the rate at which the electroactive species reach the surface (i.e., a facile electron transfer), the current is said to be *mass-transport-limited*. Such reactions are called *nernstian* or *reversible*, because they obey thermodynamic relationships. Several important techniques (discussed in Chapter 4) rely on such mass-transport-limited conditions.

1.2.1 Mass-Transport-Controlled Reactions

Mass transport occurs by three different modes:

- *Diffusion*—the spontaneous movement under the influence of concentration gradient, from regions of high concentrations to regions of lower ones, aimed at minimizing concentration differences.
- *Convection*—transport to the electrode by a gross physical movement; the major driving force for convection is an external mechanical energy associated with stirring or flowing the solution or rotating or vibrating the

electrode (i.e., forced convection). Convection can also occur naturally as a result of density gradients.

- *Migration*—movement of charged particles along an electrical field (i.e., where the charge is carried through the solution by ions according to their transference number).

These modes of mass transport are illustrated in Figure 1.1.

The *flux* (J), a common measure of the rate of mass transport at a fixed point, is defined as the number of molecules penetrating a unit area of an imaginary plane in a unit of time and is expressed in units of mol cm^{-2} s^{-1}. The flux to the electrode is described mathematically by a differential equation, known as the *Nernst–Planck equation*, given here for one dimension

$$J(x,t) = -D\frac{\partial C(x,t)}{\partial x} - \frac{zFDC}{RT}\frac{\partial \phi(x,t)}{\partial x} + C(x,t)V(x,t) \tag{1.3}$$

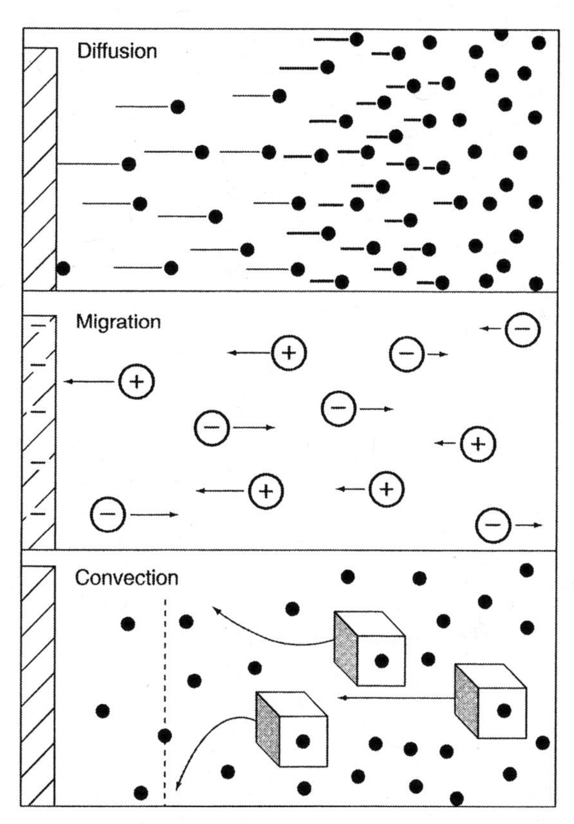

Figure 1.1 The three modes of mass transport. (Reproduced with permission from Ref. 1.)

where D is the diffusion coefficient (cm^2/s); $[\partial C(x,t)]/\partial x$ is the concentration gradient (at distance x and time t); $[\partial \phi(x,t)]/\partial x$ is the potential gradient; z and C are the charge and concentration, respectively, of the electroactive species; and $V(x,t)$ is the hydrodynamic velocity (in the x direction). In aqueous media, D usually ranges between 10^{-5} and 10^{-6} cm^2/s. The current (i) is directly proportional to the flux and the surface area (A):

$$i = -nFAJ \tag{1.4}$$

As indicated by Eq. (1.3), the situation is quite complex when the three modes of mass transport occur simultaneously. This complication makes it difficult to relate the current to the analyte concentration. The situation can be greatly simplified by suppressing the electromigration through the addition of excess inert salt. This addition of a high concentration of the supporting electrolyte (compared to the concentration of electroactive ions) helps reduce the electrical field by increasing the solution conductivity. Convection effects can be eliminated by using a quiescent solution. In the absence of migration and convection effects, movement of the electroactive species is limited by diffusion. The reaction occurring at the surface of the electrode generates a concentration gradient adjacent to the surface, which in turn gives rise to a diffusional flux. Equations governing diffusion processes are thus relevant to many electroanalytical procedures.

According to Fick's first law, the rate of diffusion (i.e., the flux) is directly proportional to the slope of the concentration gradient:

$$J(x,t) = -D\frac{\partial C(x,t)}{\partial x} \tag{1.5}$$

Combination of Eqs. (1.4) and (1.5) yields a general expression for the current response:

$$i = nFAD\frac{\partial C(x,t)}{\partial x} \tag{1.6}$$

Hence, the current (at any time) is proportional to the concentration gradient of the electroactive species. As indicated by the equations above, the diffusional flux is time-dependent. Such dependence is described by Fick's second law (for linear diffusion):

$$\frac{\partial C(x,t)}{\partial t} = D\frac{\partial^2 C(x,t)}{\partial x^2} \tag{1.7}$$

This equation reflects the rate of change with time of the concentration between parallel planes at points x and $(x + dx)$ (which is equal to the difference in flux at the two planes). Fick's second law is valid for the conditions assumed, namely, planes parallel to one another and perpendicular to the direction of diffusion, specifically, conditions of linear diffusion. In contrast, for the case of diffusion toward a spherical electrode (where the lines of flux are not parallel but are perpendicular to segments of the sphere), Fick's second law is expressed as

$$\frac{\partial C}{\partial t} = D\left[\frac{\partial^2 C}{\partial r^2} + \frac{2}{r}\frac{\partial C}{\partial r}\right] \tag{1.8}$$

where r is the distance from the center of the electrode. Overall, Fick's laws describe the flux and the concentration of the electroactive species as functions of position and time. The solution of these partial differential equations usually requires application of a (Laplace transformation) mathematical method. The Laplace transformation is of great value for such application, as it enables the conversion of the problem into a domain where a simpler mathematical manipulation is possible. Details of using the Laplace transformation are beyond the scope of this text, and can be found in Ref. 2. The establishment of proper initial and boundary conditions (which depend on the specific experiment) is also essential for this treatment. The current–concentration–time relationships resulting from such treatment are described below for several relevant experiments.

1.2.1.1 Potential-Step Experiment Let us see, for example, what happens in a potential-step experiment involving the reduction of O to R, a potential value corresponding to complete reduction of O, a quiescent solution, and a planar electrode embedded in a planar insulator. (Only O is initially present in solution.) The current–time relationship during such an experiment can be understood from the resulting concentration–time profiles. Since the surface concentration of O is zero at the new potential, a concentration gradient is established near the surface. The region within which the solution is depleted of O is known as the *diffusion layer*, and its thickness is given by δ. The concentration gradient is steep at first, and the diffusion layer is thin (see Fig. 1.2 for t_1). As time goes by, the diffusion layer expands (to δ_2 and δ_3 at t_2 and t_3), and hence the concentration gradient decreases.

Initial and boundary conditions in such an experiment include $C_O(x,0) = C_O(b)$ [i.e., at $t = 0$, the concentration is uniform throughout the system and equal to the bulk concentration, $C_O(b)$], $C_O(0,t) = 0$ for $t > 0$ (i.e., at later times the surface concentration is zero); and $C_O(x,0) \rightarrow C_O(b)$ as $x \rightarrow \infty$ (i.e., the concentration increases as the distance from the electrode increases). Solution to Fick's laws (for linear diffusion, i.e., a planar electrode) for these conditions results in a time-dependent concentration profile:

$$C_O(x,t) = C_O(b)\left\{1 - \mathrm{erf}\left[x/(4D_O t)^{1/2}\right]\right\} \tag{1.9}$$

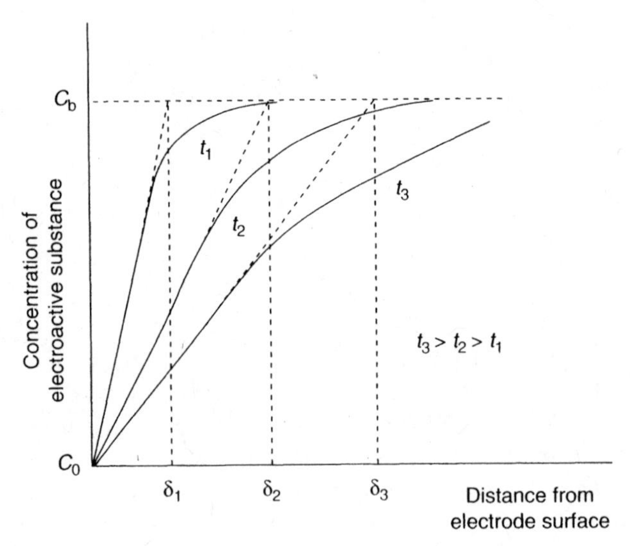

Figure 1.2 Concentration profiles for different times after the start of a potential-step experiment.

whose derivative with respect to x gives the concentration gradient at the surface

$$\frac{\partial C}{\partial x} = C_O(b)/(\pi D_O t)^{1/2} \tag{1.10}$$

when substituted into Eq. (1.6) leads to the well-known *Cottrell equation*:

$$i(t) = nFAD_O C_O(b)/(\pi D_O t)^{1/2} \tag{1.11}$$

Thus, the current decreases in proportion to the square root of time, with $(\pi D_O t)^{1/2}$ corresponding to the diffusion-layer thickness.

Solving Eq. (1.8) (using Laplace transform techniques) will yield the time evolution of the current of a spherical electrode:

$$i(t) = nFAD_O C_O(b)/(\pi D_O t)^{1/2} + nFAD_O C_O/r \tag{1.12}$$

The current response of a spherical electrode following a potential step thus contains both time-dependent and time-independent terms—reflecting the planar and spherical diffusional fields, respectively (Fig. 1.3)—becoming time independent at long timescales. As expected from Eq. (1.12), the change from one regime to another is strongly dependent on the radius of the electrode.

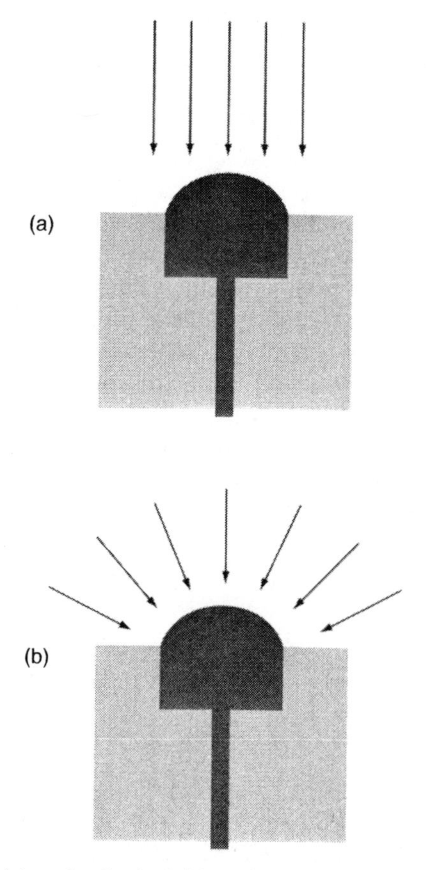

Figure 1.3 Planar (a) and spherical (b) diffusional fields at spherical electrodes.

The unique mass transport properties of ultramicroelectrodes (discussed in Section 4.5.4) are attributed to shrinkage of the electrode radius.

1.2.1.2 Potential-Sweep Experiments Let us move to a voltammetric experiment involving a linear potential scan, the reduction of O to R and a quiescent solution. The slope of the concentration gradient is given by $(C_O(b,t) - C_O(0,t))/\delta$, where $C_O(b,t)$ and $C_O(0,t)$ are the bulk and surface concentrations of O. The change in the slope, and hence the resulting current, are due to changes of both $C_O(0,t)$ and δ. First, as the potential is scanned negatively, and approaches the standard potential ($E°$) of the couple, the surface concentration rapidly decreases in accordance with the Nernst equation [Eq. (1.2)]. For example, at a potential equal to $E°$ the concentration ratio is unity $[C_O(0,t)/C_R(0,t) = 1]$. For a potential 59 mV more negative than $E°$, $C_R(0,t)$ is present at 10-fold excess $[C_O(0,t)/C_R(0,t)] = \frac{1}{10} (n = 1)$. The decrease in $C_O(0,t)$ is coupled with an increase in the diffusion-layer thickness, which dominates the change in slope after $C_O(0,t)$ approaches zero. The net result is a peak-shaped voltammogram. Such current–potential curves and the corresponding concentration–distance profiles (for selected potentials along the scan) are shown in Figure 1.4. As will be discussed in Section 4.5.4, shrinking the electrode dimension to the micrometer domain results in a sigmoid-shaped voltammetric response under quiescent conditions, characteristic of the different (radial) diffusional field and higher flux of electroactive species of ultramicroelectrodes.

Let us see now what happens in a similar linear scan voltammetric experiment, but utilizing a stirred solution. Under these conditions, the bulk concentration ($C_{O(b,t)}$) is maintained at a distance δ by the stirring. It is not influenced by the surface electron transfer reaction (as long as the electrode-area : solution-volume ratio is small). The slope of the concentration–distance profile $\{[C_O(b,t) - C_O(0,t)]/\delta\}$ is thus determined solely by the change in the surface concentration $[C_O(0,t)]$. Hence, the decrease in $C_O(0,t)$ during the potential scan (around $E°$) results in a sharp rise in the current. When a potential more negative than $E°$ by 118 mV is reached, $C_O(0,t)$ approaches zero, and a limiting current (i_l) is achieved:

$$i_l = \frac{nFAD_O C_O(b,t)}{\delta} \tag{1.13}$$

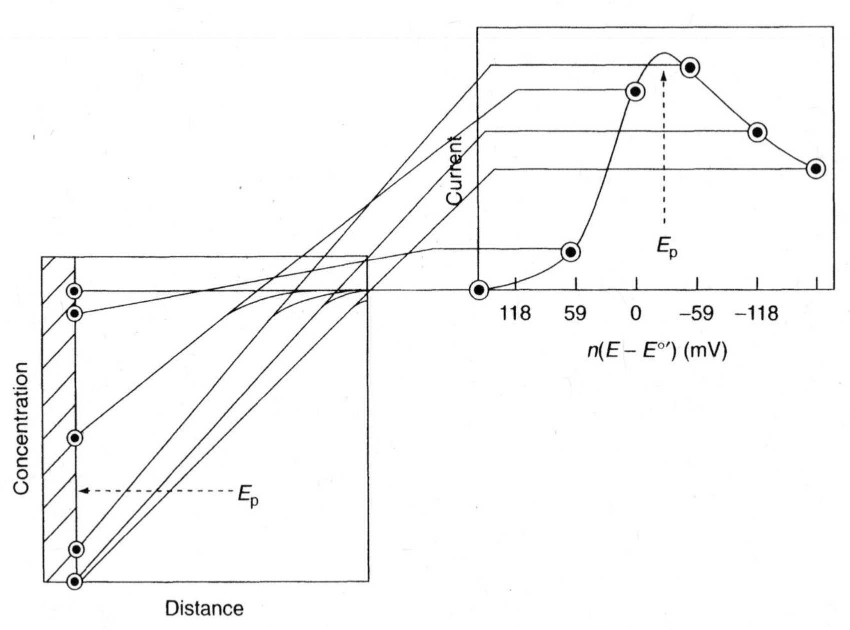

Figure 1.4 Concentration profiles (left) for different potentials during a linear sweep voltammetric experiment in unstirred solution. The resulting voltammogram is shown on the right, along with the points corresponding to each concentration gradient. (Reproduced with permission from Ref. 1.)

The resulting voltammogram thus has a sigmoidal (wave) shape. By increasing the stirring rate (U), the diffusion layer thickness becomes thinner, according to

$$\delta = \frac{B}{U^{\alpha}} \tag{1.14}$$

where B and α are constants for a given system. As a result, the concentration gradient becomes steeper (see Fig. 1.5, curve b), thereby increasing the limiting current. Similar considerations apply to other forced convection systems, including those relying on solution flow or electrode rotation (see Sections 3.6 and 4.5, respectively). For all of these hydrodynamic systems, the sensitivity of the measurement can be enhanced by increasing the convection rate.

Initially it was assumed that no solution movement occurs within the diffusion layer. Actually, a velocity gradient exists in a layer, termed the *hydrodynamic boundary layer* (or the *Prandtl layer*), where the fluid velocity increases from zero at the interface to the constant bulk value (U). The thickness of the hydrodynamic layer δ_H is related to that of the diffusion layer

$$\delta \cong \left(\frac{D}{\nu}\right)^{1/3} \delta_H \tag{1.15}$$

where ν is kinematic viscosity. In aqueous media (with $\nu \sim 10^{-2}\,cm^2/s$ and $D \sim 10^{-5}\,cm^2/s$), δ_H is approximately 10-fold larger than δ, indicating negligible con-

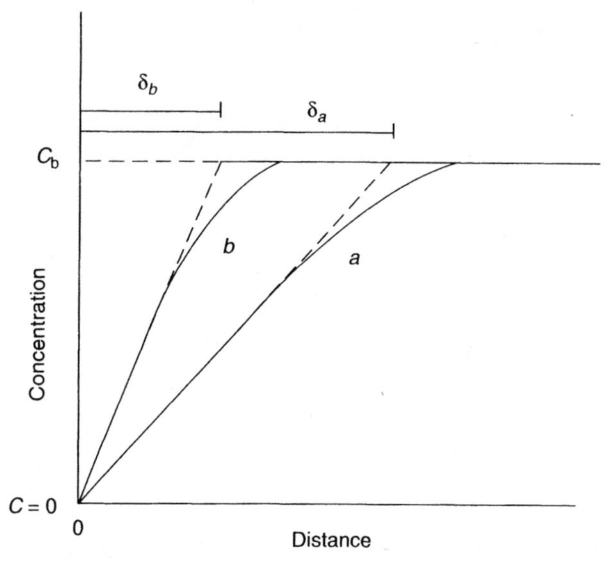

Figure 1.5 Concentration profiles for two rates of convection transport: low (curve a) and high (curve b).

vection within the diffusion layer. The discussion above applies to other forced convection systems, such as flow detectors or rotating electrodes (see Sections 3.6 and 4.5, respectively). Layer thickness (δ) values of 10–50 μm and 100–150 μm are common for electrode rotation and solution stirring, respectively. Additional means for enhancing the mass transport and thinning the diffusion layer, including the use of power ultrasound, heated electrodes, or laser activation, are also being studied (3,4). These methods may simultaneously minimize surface fouling effects, as desired for retaining the surface reactivity.

1.2.2 Reactions Controlled by the Rate of Electron Transfer

In this section we consider experiments in which the current is controlled by the rate of electron transfer (i.e., reactions with sufficiently fast mass transport). The current–potential relationship for such reactions is different from those discussed (above) for mass-transport-controlled reactions.

Consider again the electron transfer reaction: $O + ne^- \rightleftharpoons R$; the actual electron transfer step involves transfer of the electron between the conduction band of the electrode and a molecular orbital of O or R (e.g., for a reduction, from the conduction band into an unoccupied orbital in O). The rate of the forward (reduction) reaction V_f is first-order in O:

$$V_f = k_f C_O(0,t) \tag{1.16}$$

while that of the reversed (oxidation) reaction V_b, is first-order in R:

$$V_b = k_b C_R(0,t) \tag{1.17}$$

where k_f and k_b are the forward and backward heterogeneous rate constants, respectively. These constants depend on the operating potential according to the following exponential relationships:

$$k_f = k^\circ \exp[-\alpha n F(E - E^\circ)/RT] \tag{1.18}$$

$$k_b = k^\circ \exp[(1-\alpha) n F(E - E^\circ)/RT] \tag{1.19}$$

where k° is the standard heterogeneous rate constant and α is the transfer coefficient. The value of k° (in cm/s) reflects the reaction between the particular reactant and the electrode material used. The value of α (between zero and unity) reflects the symmetry of the free-energy curve (with respect to the reactants and products). For symmetric curves, α will be close to 0.5; α is a measure of the fraction of energy that is put into the system used to actually lower the activation energy (see discussion in Section 1.2.2.1). Overall, Eqs. (1.18) and (1.19) indicate that by changing the applied potential, we influence k_f and k_b in an exponential fashion. Positive and negative potentials thus speed up the oxidation and reduction reactions, respectively. For an oxidation, the energy of the electrons in the donor orbital of R must be equal to or higher than the energy of electrons in the electrode. For reduction, the energy of the electrons in the electrode must be higher than their energy in the receptor orbital of R.

Since the net reaction rate is

$$V_{net} = V_f - V_b = k_f C_O(0,t) - k_b C_R(0,t) \tag{1.20}$$

and as the forward and backward currents are proportional to V_f and V_b, respectively

$$i_f = nFAV_f \tag{1.21}$$

$$i_b = nFAV_b \tag{1.22}$$

the overall current is given by the difference between the currents due to the forward and backward reactions:

$$i_{net} = i_f - i_b = nFA[k_f C_O(0,t) - k_b C_R(0,t)] \tag{1.23}$$

By substituting the expressions for k_f and k_b [Eqs. (1.17) and (1.18), respectively], one obtains the *Butler–Volmer equation*:

$$\begin{aligned} i = nFAk°\{&C_O(0,t)\exp[-\alpha nF(E - E°)/RT] \\ &- C_R(0,t)\exp[(1-\alpha)nF(E - E°)/RT]\} \end{aligned} \tag{1.24}$$

which describes the current–potential relationship for reactions controlled by the rate of electron transfer. Note that the net current depends on both the operating potential and the surface concentration of each form of the redox (reduction–oxidation) couple. For example, Figure 1.6 displays the current–potential dependence for the case where $C_O(0,t) = C_R(0,t)$ and $\alpha = 0.50$. Large negative potentials accelerate the movement of charge in the cathodic direction, and also decelerate the charge movement in the opposite direction. As a result, the anodic current component becomes negligible and the net current merges with the cathodic component. The acceleration and deceleration of the cathodic and anodic currents are not necessarily as symmetric (as depicted in Fig. 1.6), and would differ for α values different from 0.5. Similarly, no cathodic current contribution is observed at sufficiently large positive potentials.

When $E = E°$, no net current is flowing. This situation, however, is dynamic with continuous movement of charge carriers in both directions and with equal opposing anodic and cathodic current components. The absolute magnitude of these components at $E°$ is the *exchange current* (i_0), which is directly proportional to the standard rate constant:

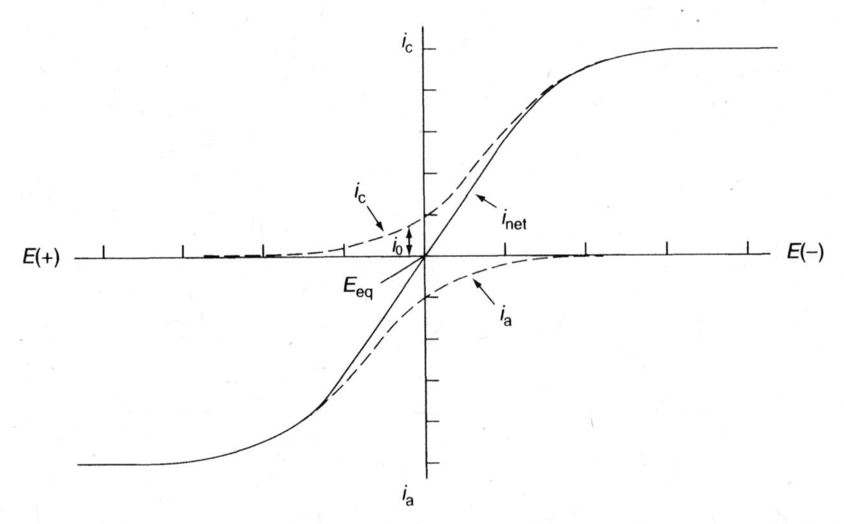

Figure 1.6 Current–potential curve for the system O + ne ↔ R, assuming that electron transfer is rate-limiting, $C_0 = C_R$, and $\alpha = 0.5$. The dotted lines show the i_c and i_a components.

$$i_0 = i_c = i_a = nFAk^\circ C \tag{1.25}$$

The exchange current density for common redox couples (at room temperature) can range from $10^{-6}\,\mu A/cm^2$ to A/cm^2. The Butler–Volmer equation can be written in terms of the exchange current

$$i = i_0[\exp(-\alpha nF\eta/RT) - \exp((1-\alpha)nF\eta/RT)] \tag{1.26}$$

where $\eta = E - E_{eq}$ is the *overvoltage* (i.e., the extra potential beyond the equilibration potential leading to a net current i). The overvoltage is always defined with respect to a specific reaction, for which the equilibrium potential is known.

Equation (1.26) can be used for extracting information on i_0 and α, which are important kinetic parameters. For sufficiently large overvoltages ($\eta > 118\,mV/n$), one of the exponential terms in Eq. (1.26) will be negligible compared with the other. For example, at large negative overpotentials, $i_c \gg i_a$ and Eq. (1.26) becomes

$$i = i_0 \exp(-\alpha nF\eta/RT) \tag{1.27}$$

and hence, we get

$$\ln i = \ln i_0 - \alpha nF\eta/RT \tag{1.28}$$

This logarithmic current–potential dependence was derived by Tafel, and is known as the *Tafel equation*. By plotting log i against η one obtains the Tafel plots for the cathodic and anodic branches of the current–overvoltage curve (Fig. 1.7). Such plots are linear only at high overpotential values; severe deviations from linearity are observed as η approaches zero. Extrapolation of the linear portions of these plots to the zero overvoltage gives an intercept, which corresponds to $\log i_0$; the slope can be used to obtain the value of the transfer coefficient α. Another form of the Tafel equation is obtained by rearrangement of Eq. (1.28):

$$\eta = a - b \log i \tag{1.29}$$

with b, the Tafel slope, having the value of $2.303RT/\alpha nF$. For $\alpha = 0.5$ and $n = 1$, this corresponds to 118 mV (at 25°C). Equation (1.29) indicates that the application of small potentials (beyond the equilibrium potential) can increase the current by many orders of magnitude. In practice, however, the current could not rise to an infinite value because of restrictions imposed by the rate at which the reactant reaches the surface. (Recall that the rate-determining step depends on the potential region.)

For small departures from $E°$, the exponential term in Eq. (1.27) may be linearized, with the current approximately proportional to η:

$$i = i_0 \, nF\eta / RT \tag{1.30}$$

Hence, the net current is directly proportional to the overvoltage in a narrow potential range near $E°$.

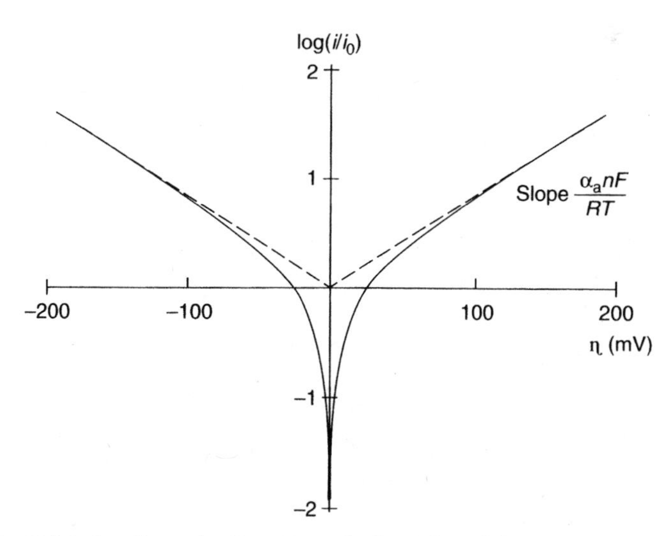

Figure 1.7 Tafel plots for cathodic and anodic branches of the current–potential curve.

Note also that at equilibrium ($E = E_{eq}$) the net current is zero (i.e., equal currents are passing reversibly in both directions); one can thus obtain the following from Eq. (1.24):

$$C_O(0,t)\exp[-\alpha nF(E-E^\circ)/RT] = C_R(0,t)\exp[(1-\alpha)nF(E-E^\circ)/RT] \quad (1.31)$$

Rearrangement of Eq. (1.31) yields the exponential form of the Nernst equation

$$\frac{C_O(0,t)}{C_R(0,t)} = \exp[nF(E-E^\circ)/RT] \quad (1.32)$$

expected for equilibrium conditions.

The equilibrium potential for a given reaction is related to the formal potential

$$E_{eq} = E^\circ + (2.3RT/nF)\log Q \quad (1.33)$$

where Q is the equilibrium ratio function (i.e., ratio of the equilibrium concentrations).

1.2.2.1 Activated Complex Theory The effect of the operating potential on the rate constants [Eqs. (1.18) and (1.19)] can be understood in terms of the free-energy barrier. Figure 1.8 shows a typical Morse potential energy

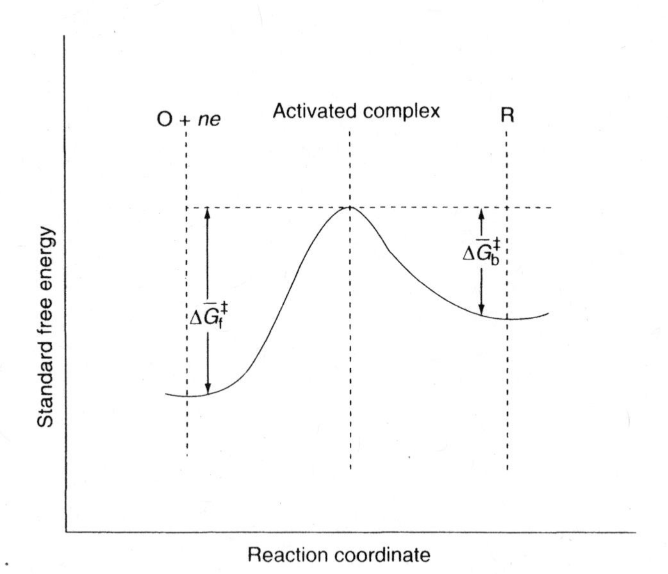

Figure 1.8 Free-energy curve for a redox process at a potential more positive than the equilibrium value.

curve for the reaction $O + ne^- \rightleftharpoons R$, at an inert metallic electrode (where O and R are soluble). Because of the somewhat different structures of O and R, there is a barrier to electron transfer (associated with changes in bond lengths and bond angles). In order for the transition from the oxidized form to occur, it is thus necessary to overcome the free energy of activation, $\Delta G^{\ddagger}$. The frequency with which the electron crosses the energy barrier as it moves from the electrode to O (i.e., the rate constant) is given by

$$k = Ae^{-\Delta G^{\ddagger}/RT} \tag{1.34}$$

Any alteration in $\Delta G^{\ddagger}$ will thus affect the rate of the reaction. If $\Delta G^{\ddagger}$ is increased, the reaction rate will decrease. At equilibrium, the cathodic and anodic activation energies are equal ($\Delta G^{\ddagger}_{c,0} = \Delta G^{\ddagger}_{a,0}$) and the probability of electron transfer will be the same in both directions. A, known as the frequency factor, is given as a simple function of the Boltzmann constant k' and the Planck constant, h:

$$A = \frac{k'T}{h} \tag{1.35}$$

Now let us discuss nonequilibirum situations. By varying the potential of the working electrode, we can influence the free energy of its resident electrons, thus making one reaction more favorable. For example, a potential shift E from the equilibrium value moves the $O + ne^-$ curve up or down by $\phi = -nFE$. The dashed line in Figure 1.9 displays such a change for the case of

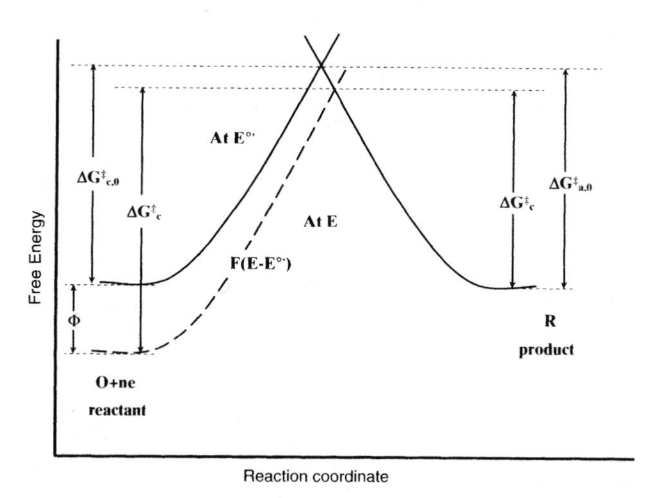

Figure 1.9 Effect of a change in the applied potential on the free energies of activation for reduction and oxidation.

a positive E. Under this condition the barrier for reduction, $\Delta G_c^\ddagger$, is larger than $\Delta G_{c,0}^\ddagger$. A careful study of the new curve reveals that only a fraction (α) of the energy shift ϕ is actually used to increase the activation energy barrier, and hence to accelerate the rate of the reaction. On the basis of the symmetry of the two potential curves, this fraction (the transfer coefficient) can range from zero to unity. Measured values of α in aqueous solutions have ranged from 0.2 to 0.8. The term α is thus a measure of the symmetry of the activation energy barrier. A α value of 0.5 indicates that the activated complex is exactly halfway between the reagents and products on the reaction coordinate (i.e., an idealized curve); α values close to 0.5 are common for metallic electrodes with a simple electron transfer process. The barrier for reduction at E is thus given by

$$\Delta G_c^\ddagger = \Delta G_{c,0}^\ddagger + \alpha nFE \tag{1.36}$$

Similarly, examination of the figure reveals also that the new barrier for oxidation, $\Delta G_a^\ddagger$, is lower than $\Delta G_{a,0}^\ddagger$:

$$\Delta G_a^\ddagger = \Delta G_{a,0}^\ddagger - (1-\alpha)nFE \tag{1.37}$$

By substituting the expressions for $\Delta G^\ddagger$ [Eqs. (1.36) and (1.37)] in Eq. (1.34), we obtain the following equations for reduction

$$k_f = A\exp[-\Delta G_{c,0}^\ddagger/RT]\cdot\exp[-\alpha nFE/RT] \tag{1.38}$$

and for oxidation:

$$k_b = A\exp[-\Delta G_{a,0}^\ddagger/RT]\cdot\exp[(1-\alpha)nFE/RT] \tag{1.39}$$

The first two factors in Eqs. (1.38) and (1.39) are independent of the potential, and thus these equations can be rewritten as

$$k_f = k_f^\circ\exp[-\alpha nFE/RT] \tag{1.40}$$

$$k_b = k_b^\circ\exp[(1-\alpha)nFE/RT] \tag{1.41}$$

When the electrode is at equilibrium with the solution, and when the surface concentrations of O and R are the same, $E = E^\circ$, and k_f and k_b are equal

$$k_f^\circ\exp[-\alpha nFE/RT] = k_b^\circ\exp[(1-\alpha)nFE/RT] = k^\circ \tag{1.42}$$

and correspond to the standard rate constant k°. By substituting for k_f° and k_b° [using Eq. (1.42)] in Eqs. (1.40) and (1.41), one obtains Eqs. (1.18) and (1.19) (which describe the effect of the operating potential on the rate constants).

1.3 ELECTRICAL DOUBLE LAYER

The electrical double layer is the array of charged particles and/or oriented dipoles existing at every material interface. In electrochemistry, such a layer reflects the ionic zones formed in the solution, to compensate for the excess of charge on the electrode (q_e). A positively charged electrode thus attracts a layer of negative ions (and vice versa). Since the interface must be neutral, $q_e + q_s = 0$ (where q_s is the charge of the ions in the nearby solution). Accordingly, such a counterlayer consists of ions of sign opposite that of the electrode. As illustrated in Figure 1.10, the electrical double layer has a complex structure of several distinct parts.

The inner layer (closest to the electrode), known as the *inner Helmholz plane* (IHP), contains solvent molecules and specifically adsorbed ions (such as Br^- or I^- that are not hydrated in aqueous solutions). It is defined by the locus of points for the specifically adsorbed ions. The next layer, the *outer Helmholz plane* (OHP), reflects the imaginary plane passing through the

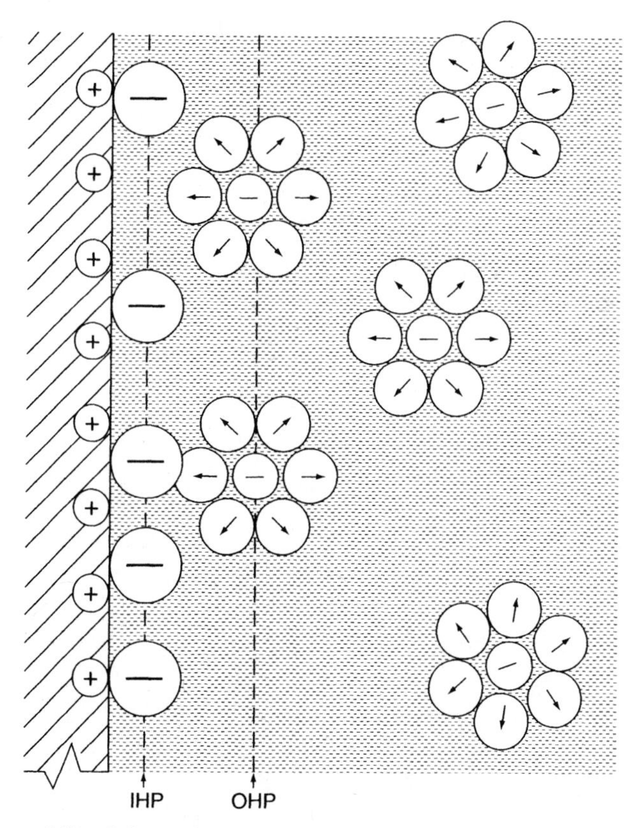

Figure 1.10 Schematic representation of the electrical double layer.

center of solvated ions at their closest approach to the surface. The solvated ions are nonspecifically adsorbed and are attracted to the surface by long-range coulombic forces. Both Helmholz layers represent the compact layer. Such a compact layer of charges is strongly held by the electrode and can survive even when the electrode is pulled out of the solution. The Helmholz model does not take into account the thermal motion of ions, which loosens them from the compact layer.

The outer layer (beyond the compact layer), referred to as the *diffuse layer* (or *Gouy layer*), is a three-dimensional region of scattered ions, which extends from the OHP into the bulk solution. Such ionic distribution reflects the counterbalance between ordering forces of the electrical field and the disorder caused by a random thermal motion. The equilibrium between these two opposing effects, indicates that the concentration of ionic species at a given distance from the surface, $C(x)$, decays exponentially with the ratio between the electrostatic energy ($zF\Phi$) and the thermal energy (RT), in accordance with the Boltzmann equation):

$$C(x) = C(0)\exp(-zF\Phi)/RT) \tag{1.43}$$

The total charge of the compact and diffuse layers equals (and is opposite in sign) to the net charge on the electrode side. The potential-distance profile across the double-layer region involves two segments, with a linear increase until the OHP and an exponential one within the diffuse layer. Such two-potential drops are displayed in Figure 1.11. Depending on the ionic strength, the thickness of the double layer may extend to more than 10 nm.

The electrical double layer resembles an ordinary (parallel-plate) capacitor. For an ideal capacitor, the charge (q) is directly proportional to the potential difference:

$$q = CE \tag{1.44}$$

where C is the capacitance (in farads, F), specifically, the ratio of the charge stored to the applied potential. The potential–charge relationship for the electrical double layer is

$$q = C_{\text{dl}}A(E - E_{\text{pzc}}) \tag{1.45}$$

where C_{dl} is the capacitance per unit area and E_{pzc} is the potential of zero charge (i.e., where the sign of the electrode charge reverses and no net charge exists in the double layer). The C_{dl} values are usually in the range of 10–40 $\mu F/cm^2$.

The capacitance of the double layer consists of a combination of the capacitance of the compact layer in series with that of the diffuse layer. As is common for two capacitors in series, the total capacitance is given by

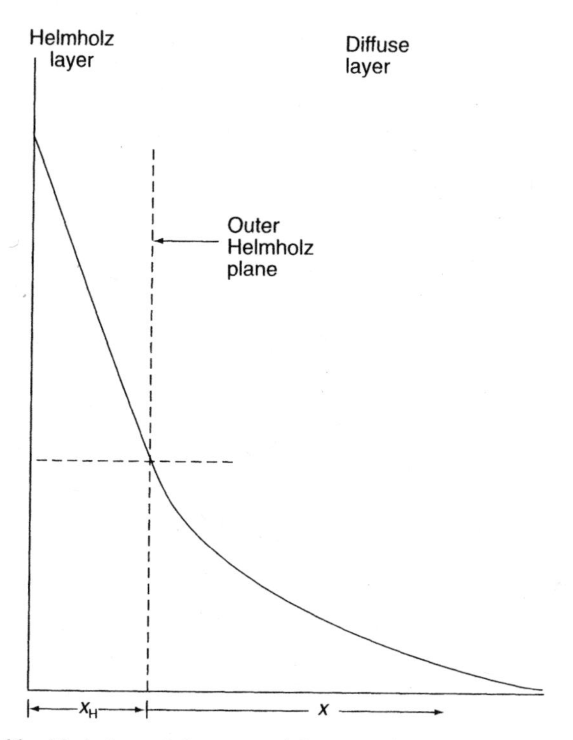

Figure 1.11 Variation of the potential across the electrical double layer.

$$1/C = 1/C_H + 1/C_G \tag{1.46}$$

where C_H and C_G represent that capacitances of the compact and diffuse layers, respectively. The smaller of these capacitances determines the observed behavior. By analogy to a parallel-plate (ideal) capacitor, C_H is given by

$$C_H = -\varepsilon/4\pi d \tag{1.47}$$

where d is the distance between the plates and ε is the dielectric constant. ($\varepsilon = 78$ for water at room temperature.) Accordingly, C_H increases with decreasing separation between the electrode surface and the counterionic layer, as well as with increasing the dielectric constant in the intervening medium. The value of C_G is strongly affected by the electrolyte concentration; the compact layer is largely independent of the concentration. For example, at sufficiently high electrolyte concentration, most of the charge is confined near the Helmholz plane, and little is scattered diffusely into the solution (i.e., the diffuse double layer becomes sufficiently small). Under these conditions, $1/C_H \gg 1/C_G$, $1/C \simeq 1/C_H$, or $C \simeq C_H$. In contrast, for dilute solutions, C_G is very small (compared to C_H) and $C \simeq C_G$.

Figure 1.12 displays the experimental dependence of the double-layer capacitance on the applied potential and electrolyte concentration. As expected for the parallel-plate model, the capacitance is nearly independent of the potential or concentration over several hundred millivolts. Yet, a sharp dip in the capacitance is observed (around −0.5 V) with dilute solutions, reflecting the contribution of the diffuse layer. The charging of the double layer is responsible for the background (residual) current known as the *charging current*, which limits the detectability of controlled-potential techniques. Such a charging process is nonfaradaic because electrons are not transferred across the electrode–solution interface. It occurs when a potential is applied across the double layer, or when the electrode area or capacitances are changing. Note that the current is the time derivative of the charge. Hence, when such processes occur, a residual current is flowing according to the differential equation

$$i = \frac{dq}{dt} = C_{\mathrm{dl}} A \frac{dE}{dt} + C_{\mathrm{dl}}(E - E_{\mathrm{pzc}})\frac{dA}{dt} + A(E - E_{\mathrm{pzc}})\frac{dC_{\mathrm{dl}}}{dt} \tag{1.48}$$

where dE/dt and dA/dt are the potential scan rate and rate of area change, respectively. The second term is applicable to the dropping mercury electrode (discussed in Section 4.2). The term dC_{dl}/dt is important when adsorption processes change the double-layer capacitance.

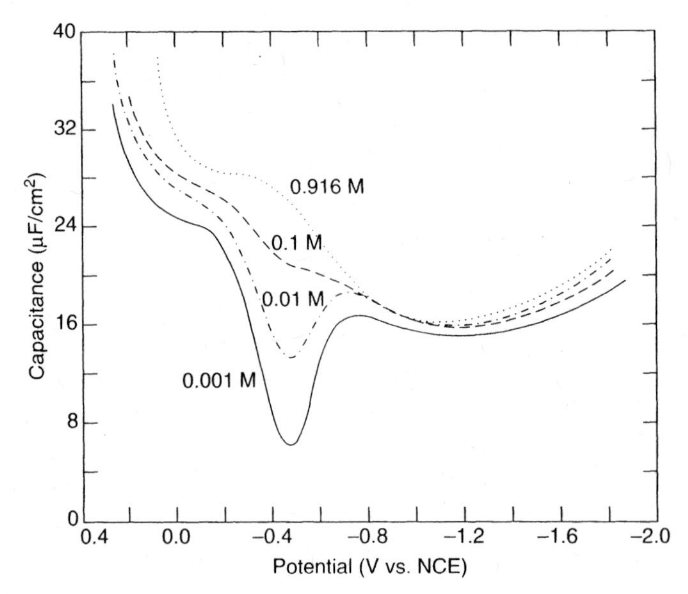

Figure 1.12 Double-layer capacitance of a mercury drop electrode in NaF solutions of different concentrations. (Reproduced with permission from Ref. 5.)

Alternately, for potential-step experiments (e.g., chronoamperometry; see Section 3.1), the charging current is the same as that obtained when a potential step is applied to a series *RC* circuit:

$$i_c = \frac{E}{R_S} e^{-t/RC_{dl}} \tag{1.49}$$

Thus, the current decreases exponentially with time. Here, E is the magnitude of the potential step, while R_s is the (uncompensated) solution resistance.

Equation (1.48) can be used for calculating the double-layer capacitance of solid electrodes. By recording linear scan voltammograms at different scan rates (using the supporting electrolyte solution), and plotting the charging current (at a given potential) versus the scan rate, one would obtain a straight line, with a slope corresponding to $C_{dl}A$.

Measurements of the double-layer capacitance provide valuable insights into adsorption and desorption processes, as well as into the structure of film-modified electrodes (6).

Further discussion of the electrical double layer can be found in several reviews (5,7–11).

1.4 ELECTROCAPILLARY EFFECT

Electrocapillary is the study of the interfacial tension as a function of the electrode potential. Such a study can shed useful light on the structure and properties of the electrical double layer. The influence of the electrode–solution potential difference on the surface tension (γ) is particularly pronounced at nonrigid electrodes (such as the dropping mercury one, discussed in Section 4.5). A plot of the surface tension versus the potential (like the ones shown in Fig. 1.13) is called an *electrocapillary curve.*

The excess charge on the electrode can be obtained from the slope of the electrocapillary curve (at any potential), by the Lippman equation:

$$\left(\frac{\partial \gamma}{\partial E}\right)_{\text{const.pressure}} = q \tag{1.50}$$

The more highly charged the interface becomes, the more the charges repel each other, thereby decreasing the cohesive forces and lowering the surface tension. The second differential of the electrocapillary plot gives directly the differential capacitance of the double layer:

$$\frac{\partial^2 \gamma}{\partial E^2} = -C_{dl} \tag{1.51}$$

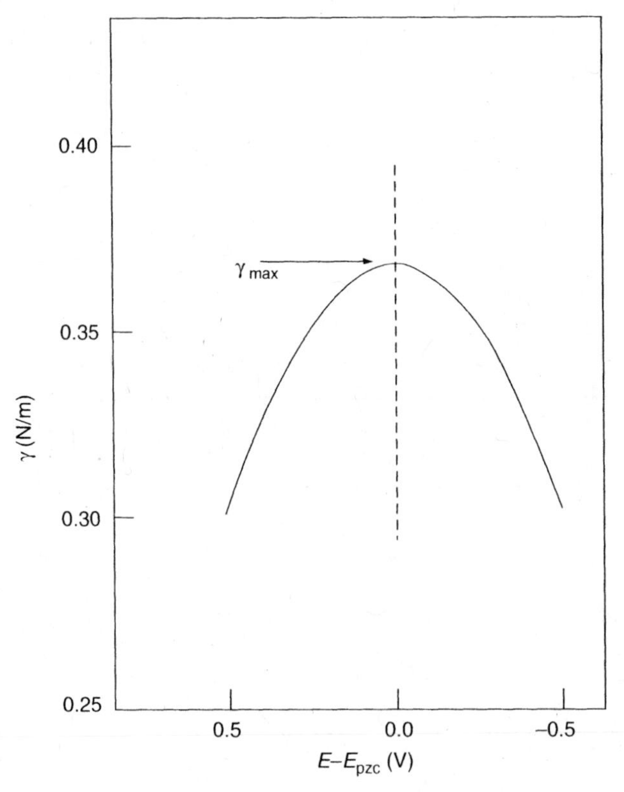

Figure 1.13 Electrocapillary curve of surface tension (γ) versus the potential.

Hence, the differential capacitance represents the slope of the plot of q versus E.

An important point of the electrocapillary curve is its maximum. Such maximum value of γ, obtained when $q = 0$, corresponds to the potential of zero charge (E_{pzc}). The surface tension is a maximum because on the uncharged surface there is no repulsion between like charges. The charge on the electrode changes its sign after passing the potential through the E_{pzc}. Experimental electrocapillary curves have a nearly parabolic shape around the E_{pzc}. Such a parabolic shape corresponds to a linear change of the charge with the potential. The deviation from a parabolic shape depends on the solution composition, particularly on the nature of the anions present in the electrolyte. In particular, specific interaction of various anions (e.g., halides) with the mercury surface, occurring at positive potentials, causes deviations from the parabolic behavior (with shifts of E_{pzc} to more cathodic potentials). As shown in Figure 1.14, the change in surface tension and the negative shift in E_{pzc} increase in the following order: $I^- > Br^- > CNS^- > NO_3^- > OH^-$. (These changes are expected from the strength of the specific adsorption.) Such ions can be specifically

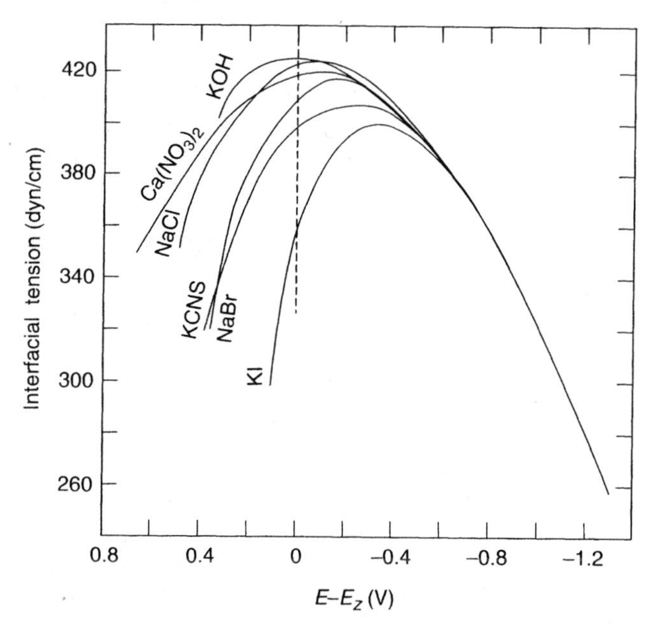

Figure 1.14 Electrocapillary curves for different electrolytes showing the relative strength of specific adsorption. (Reproduced with permission from Ref. 7.)

adsorbed because they are not solvated. Inorganic cations, in contrast, are less specifically adsorbed (because they are usually hydrated). Similarly, blockage of the surface by a neutral adsorbate often causes depressions in the surface tension in the vicinity of the E_{pzc} (Fig. 1.15). Note the reduced dependence of γ on the potential around this potential. At more positive or negative potentials, such adsorbates are displaced from the surface by oriented water molecules.

1.5 SUPPLEMENTARY READING

Several international journals bring together papers and reviews covering innovations and trends in the field of electroanalytical chemistry:

Bioelectrochemistry and Bioenergetics
Biosensors and Bioelectronics
Electroanalysis
Electrochemistry Communications
Electrochimica Acta
Journal of Applied Electrochemistry
Journal of Electroanalytical and Interfacial Electrochemistry

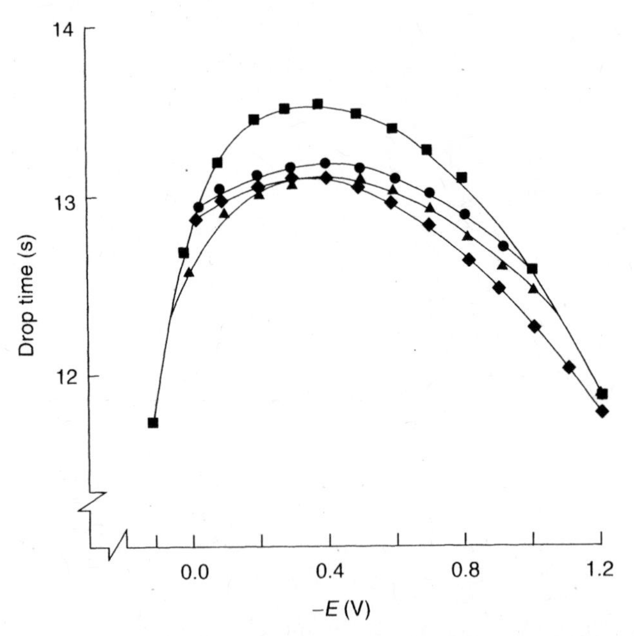

Figure 1.15 Electrocapillary curves of background (■), ethynylestradiol (●), β-estradion (△), and morgestrel (◆). (Reproduced with permission from Ref. 12.)

Journal of the Electrochemical Society
Langmuir
Sensors and Actuators

Useful information can be found in many prominent journals that cater to all branches of analytical chemistry, including *The Analyst*, *Analytica Chimica Acta*, *Analytical Chemistry*, *Talanta*, *Analytical Letters*, and *Analytical and Bioanalytical Chemistry*. Biennial reviews published in the June issue of *Analytical Chemistry* offer comprehensive summaries of fundamental and practical research work.

Many textbooks and reference works dealing with various aspects of electroanalytical chemistry have been published since the 1960s. Some of these are listed below as suggestions for additional reading, in alphabetic order:

Albery, W. J., *Electrode Kinetics*, Clarendon Press, Oxford, UK, 1975.

Bard, A. J.; Faulkner, L., *Electrochemical Methods*, 2nd ed., Wiley, New York, 2000.

Bond, A. M., *Modern Polarographic Methods in Analytical Chemistry*, Marcel Dekker, New York, 1980.

Bockris, J. M.; Reddy, A., *Modern Electrochemistry*, Vols. 1, 2, Plenum Press, New York, 1970.

Brett, C.; Oliveira Brett, A. M., *Electrochemistry: Principles, Methods and Applications*, Oxford Univ. Press, Oxford, UK, 1993.

Diamond, D., *Chemical and Biological Sensors*, Wiley, New York, 1998.

Gileadi, E., *Electrode Kinetics*, VCH Publishers, New York, 1993.

Kissinger, P.; Heineman, W., *Laboratory Techniques in Electroanalytical Chemisry*, 2nd ed., Marcel Dekker, New York, 1996.

Janata, J., *Principles of Chemical Sensors*, Plenum Press, New York, 1989.

Koryta, J.; Dvorak, J., *Principles of Electrochemistry*, Wiley, Chichester, UK, 1987.

Rieger, P., *Electrochemistry*, Prentice-Hall, Englewood Cliffs, NJ, 1987.

Sawyer, D.; Roberts, J., *Experimental Electrochemistry for Chemists*, Wiley, New York, 1974.

Smyth, M.; Vos, J., *Analytical Voltammetry*, Elsevier, Amsterdam, 1992.

Turner, A. P.; Karube, I.; Wilson, G., *Biosensors*, Oxford Univ. Press, Oxford, UK, 1987.

Wang, J., *Electroanalytical Techniques in Clinical Chemistry and Laboratory Medicine*, VCH Publishers, New York, 1988.

PROBLEMS

1.1 Describe and draw the concentration profile or gradient near the electrode surface during a linear scan voltammetric experiment in a *stirred* solution. (Use five or six potentials on both sides of $E°$.) Show also the resulting voltammogram, along with points for each concentration gradient (in a manner analogous to Fig. 1.4).

1.2 Describe and draw the structure of the electrical double layer (with its several distinct parts).

1.3 Use the activated complex theory for explaining how the applied potential affects the rate constant of an electron transfer reaction. Use or draw free-energy curves and use proper equations for your explanation.

1.4 Use equations to demonstrate how an increase in the stirring rate will affect the mass-transport-controlled limiting current.

1.5 Derive the Nernst equation from the Butler–Volmer equation.

1.6 Explain clearly why polyanionic DNA molecules adsorb onto electrode surfaces at potentials more positive than E_{pzc}, and suggest a protocol for desorbing them back to the solution.

1.7 Which experimental conditions assure that the movement of the electroactive species is limited by diffusion? How do these conditions suppress the migration and convection effects?

1.8 Explain clearly the reason for the peaked response of linear sweep voltammetric experiments involving a planar macrodisk electrode and a quiescent solution.

1.9 The net current flowing at the equilibrium potential is zero, yet this is a dynamic situation with equal opposing cathodic and anodic current components (whose absolute value is i_0). Suggest an experimental route for estimating the value of i_0.

1.10 Explain clearly why only a fraction of the energy shift (associated with a potential shift) is used for increasing the activation energy barrier.

REFERENCES

1. Maloy, J. R., *J. Chem. Educ.* **60**, 285 (1983).
2. Smith, M. G., *Laplace Transform Theory*, Van Nostrand, London, 1966.
3. Compton, R. G.; Eklund, J.; Marken, F., *Electroanalysis* **9**, 509 (1997).
4. Grundler, P.; Kirbs, A., *Electroanalysis* **11**, 223 (1999).
5. Grahame, D., *Chem. Rev.* **41**, 441 (1947).
6. Swietlow, A.; Skoog, M.; Johansson, G., *Electroanalysis* **4**, 921 (1992).
7. Grahame, D. C., *Ann. Rev. Phys. Chem.* **6**, 337 (1955).
8. Mohilner, D., *J. Electroanal. Chem.* **1**, 241 (1966).
9. Bockris, O'M.; Devanathan, M. A.; Muller, K., *Proc. Roy. Soc. Lond.* **55**, A274 (1963).
10. Parsons, R., *J. Electrochem. Soc.* **127**, 176C (1980).
11. Mark, H. B., *Analyst* **115**, 667 (1990).
12. Bond, A. M.; Heritage, I.; Briggs, M., *Langmuir* **1**, 110 (1985).

2

STUDY OF ELECTRODE REACTIONS AND INTERFACIAL PROPERTIES

2.1 CYCLIC VOLTAMMETRY

Cyclic voltammetry is the most widely used technique for acquiring qualitative information about electrochemical reactions. The power of cyclic voltammetry results from its ability to rapidly provide considerable information on the thermodynamics of redox processes and the kinetics of heterogeneous electron transfer reactions and on coupled chemical reactions or adsorption processes. Cyclic voltammetry is often the first experiment performed in an electroanalytical study. In particular, it offers a rapid location of redox potentials of the electroactive species, and convenient evaluation of the effect of media on the redox process.

Cyclic voltammetry consists of scanning linearly the potential of a stationary working electrode (in an unstirred solution), using a triangular potential waveform (Fig. 2.1). Depending on the information sought, single or multiple cycles can be used. During the potential sweep, the potentiostat measures the current resulting from the applied potential. The resulting current–potential plot is termed a *cyclic voltammogram*. The cyclic voltammogram is a complicated, time-dependent function of a large number of physical and chemical parameters.

Figure 2.2 illustrates the expected response of a reversible redox couple during a single potential cycle. It is assumed that only the oxidized form O is present initially. Thus, a negative-going potential scan is chosen for the first half-cycle, starting from a value where no reduction occurs. As the applied

Analytical Electrochemistry, Third Edition, by Joseph Wang

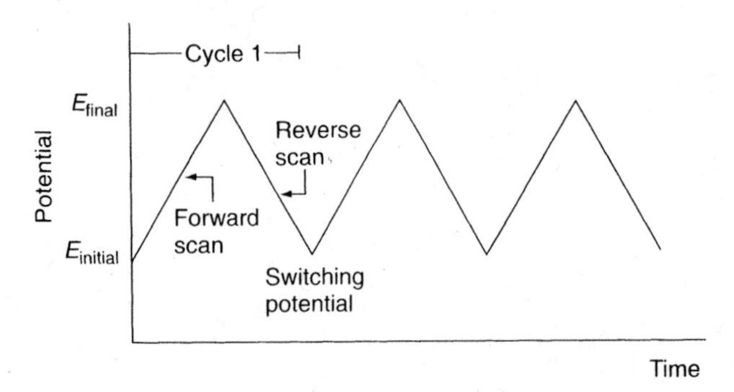

Figure 2.1 Potential–time excitation signal in a cyclic voltammetric experiment.

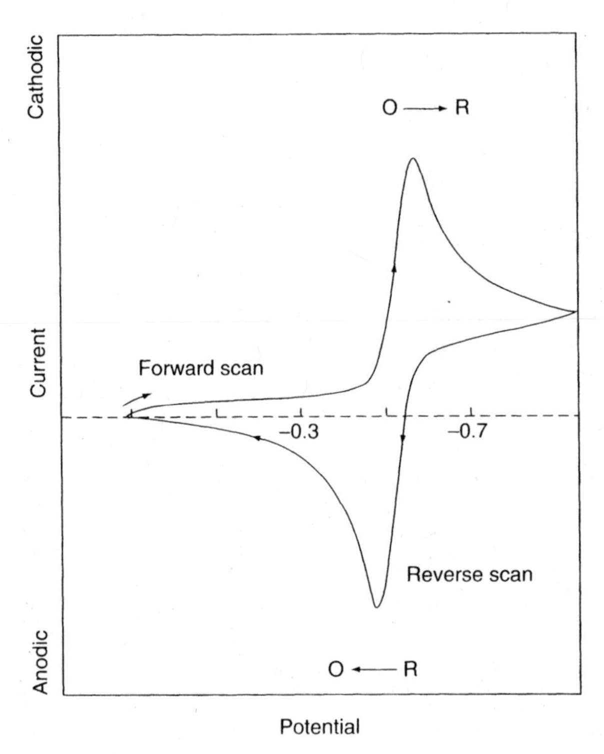

Figure 2.2 Typical cyclic voltammogram for a reversible O + ne^- ⇋ R redox process.

potential approaches the characteristic E° for the redox process, a cathodic current begins to increase, until a peak is reached. After traversing the potential region in which the reduction process takes place (at least $90/n$ mV beyond the peak), the direction of the potential sweep is reversed. During the reverse scan, R molecules (generated in the forward half-cycle, and accumulated near the surface) are reoxidized back to O, resulting in an anodic peak.

The characteristic peaks in the cycle voltammogram are caused by the formation of the diffusion layer near the electrode surface. These can be best understood by carefully examining the concentration–distance profiles during the potential sweep (see Section 1.2.1.2). For example, Figure 2.3 illustrates four concentration gradients for the reactant and product at different times corresponding to (a) the initial potential value, (b,c) the formal potential of the couple (during the forward and reversed scans, respectively), and (c) the achievement of a zero-reactant surface concentration. Note that the continuous change in the surface concentration is coupled with an expansion of the diffusion-layer thickness (as expected in quiescent solutions). The resulting current peaks thus reflect the continuous change of the concentration gradient with time. Hence, the increase in the peak current corresponds to the achievement of diffusion control, while the current drop (beyond the peak)

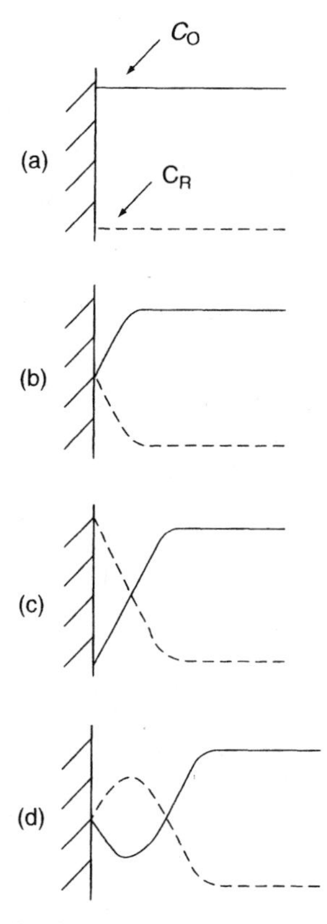

Figure 2.3 Concentration distribution of the oxidized and reduced forms of the redox couple at different times during a cyclic voltammetric experiment corresponding to the initial potential (a), to the formal potential of the couple during the forward and reversed scans (b,d), and to the achievement of a zero-reactant surface concentration (c).

exhibits a $t^{-1/2}$ dependence (independent of the applied potential). For these reasons, the reversal current has the same shape as does the forward one. As will be discussed in Chapter 4, the use of ultramicroelectrodes—for which the mass transport process is dominated by radial (rather than linear) diffusion—results in a sigmoid-shaped cyclic voltammogram.

2.1.1 Data Interpretation

The cyclic voltammogram is characterized by several important parameters. Four of these observables, the two peak currents and two peak potentials, provide the basis for the diagnostics developed by Nicholson and Shain (1) for analyzing the cyclic voltammetric response.

2.1.1.1 Reversible Systems The peak current for a reversible couple (at 25°C) is given by the Randles–Sevcik equation

$$i_p = (2.69 \times 10^5) n^{3/2} A C D^{1/2} v^{1/2} \tag{2.1}$$

where n is the number of electrons, A the electrode area (in cm^2), C the concentration (in mol/cm^3), D the diffusion coefficient (in cm^2/s), and v the potential scan rate (in V/s). Accordingly, the current is directly proportional to concentration and increases with the square root of the scan rate. Such dependence on the scan rate is indicative of electrode reaction controlled by mass transport (semiinfinite linear diffusion). The reverse-to-forward peak current ratio, $i_{p,r}/i_{p,f}$, is unity for a simple reversible couple. As will be discussed in the following sections, this peak ratio can be strongly affected by chemical reactions coupled to the redox process. The current peaks are commonly measured by extrapolating the preceding baseline current.

The position of the peaks on the potential axis (E_p) is related to the formal potential of the redox process. The formal potential for a reversible couple is centered between $E_{p,a}$ and $E_{p,c}$:

$$E° = \frac{E_{p,a} + E_{p,c}}{2} \tag{2.2}$$

The separation between the peak potentials (for a reversible couple) is given by

$$\Delta E_p = E_{p,a} - E_{p,c} = \frac{0.059}{n} \text{ V} \tag{2.3}$$

Thus, the peak separation can be used to determine the number of electrons transferred, and as a criterion for a Nernstian behavior. Accordingly, a fast one-electron process exhibits a ΔE_p of about 59 mV. Both the cathodic and anodic peak potentials are independent of the scan rate. It is possible to relate the half-peak potential ($E_{p/2}$, where the current is half of the peak current) to the polarographic half-wave potential, $E_{1/2}$:

$$E_{p/2} = E_{1/2} \pm \frac{0.028}{n} \quad \text{V} \tag{2.4}$$

(The sign is positive for a reduction process.)

For multielectron transfer (reversible) processes, the cyclic voltammogram consists of several distinct peaks, if the E° values for the individual steps are successively higher and are well separated. An example of such a mechanism is the six-step reduction of the fullerenes C_{60} and C_{70} to yield the hexaanion products C_{60}^{6-} and C_{70}^{6-}. Such six successive reduction peaks are observed in Figure 2.4.

The situation is very different when the redox reaction is slow or coupled with a chemical reaction. Indeed, it is these "nonideal" processes that are usually of greatest chemical interest and for which the diagnostic power of

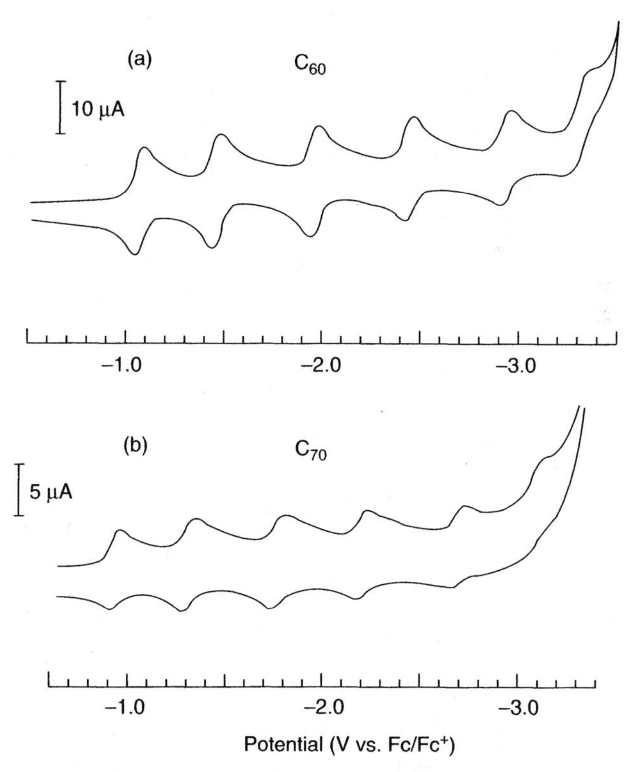

Figure 2.4 Cyclic voltammetry of C_{60} and C_{70} in an acetonitrile/toluene solution. (Reproduced with permission from Ref. 2.)

cyclic voltammetry is most useful. Such information is usually obtained by comparing the experimental voltammograms with those derived from theoretical (simulated) ones (1). Proper compensation of the ohmic drop (see Section 4.4) is crucial for such diagnostic applications of cyclic voltammetry.

2.1.1.2 Irreversible and Quasi-reversible Systems For irreversible processes (those with sluggish electron exchange), the individual peaks are reduced in size and widely separated (Fig. 2.5, curve *A*). Totally irreversible systems are characterized by a shift of the peak potential with the scan rate:

$$E_p = E^\circ - \frac{RT}{\alpha n_a F}\left[0.78 - \ln\frac{k^\circ}{D^{1/2}} + \ln\left(\frac{\alpha n_a F v}{RT}\right)^{1/2}\right] \tag{2.5}$$

where α is the transfer coefficient and n_a is the number of electrons involved in the charge transfer step. Thus, E_p occurs at potentials higher than E°, with the overpotential related to k° and α. Independent of the value k°, such peak displacement can be compensated by an appropriate change of the scan rate. The peak potential and the half-peak potential (at 25°C) will differ by $48/\alpha n$ mV. Hence, the voltammogram becomes more drawn-out as αn decreases.

The peak current, given by

$$i_p = (2.99 \times 10^5) n (\alpha n_a)^{1/2} ACD^{1/2} v^{1/2} \tag{2.6}$$

is still proportional to the bulk concentration, but will be lower in height (depending on the value of α. Assuming an value of 0.5, the ratio of the reversible-to-irreversible current peaks is 1.27 (i.e., the peak current for the irreversible process is about 80% of the peak for a reversible one).

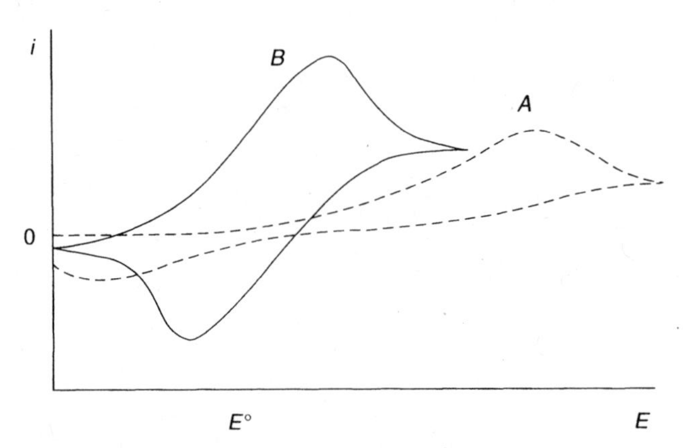

Figure 2.5 Cyclic voltammograms for irreversible (curve *A*) and quasi-reversible (curve *B*) redox processes.

For quasi-reversible systems (with $10^{-1} > k^\circ > 10^{-5}$ cm/s) the current is controlled by both the charge transfer and mass transport. The shape of the cyclic voltammogram is a function of $k^\circ/\sqrt{\pi a D}$ (where $a = nFv/RT$). As $k^\circ/\sqrt{\pi a D}$ increases, the process approaches the reversible case. For small values of $k^\circ/\sqrt{\pi a D}$ (i.e., at very fast v), the system exhibits an irreversible behavior. Overall, the voltammograms of a quasi-reversible system are more drawn out and exhibit a larger separation in peak potentials compared to a reversible system (Fig. 2.5, curve *B*).

2.1.2 Study of Reaction Mechanisms

One of the most important applications of cyclic voltammetry is for qualitative diagnosis of chemical reactions that precede or succeed the redox process (1). Such reaction mechanisms are commonly classified by using the letters E and C (for the redox and chemical steps, respectively) in the order of the steps in the reaction scheme. The occurrence of such chemical reactions, which directly affect the available surface concentration of the electroactive species, is common to redox processes of many important organic and inorganic compounds. Changes in the shape of the cyclic voltammogram, resulting from the chemical competition for the electrochemical reactant or product, can be extremely useful for elucidating these reaction pathways and for providing reliable chemical information about reactive intermediates.

For example, when the redox system is perturbed by a following chemical reaction, namely, an EC mechanism

$$\mathrm{O} + ne^- \rightleftharpoons \mathrm{R} \rightarrow \mathrm{Z} \tag{2.7}$$

the cyclic voltammogram will exhibit a smaller reverse peak (because the product R is chemically 'removed' from the surface). The peak ratio $i_{p,r}/i_{p,f}$ will thus be smaller than unity; the exact value of the peak ratio can be used to estimate the rate constant of the chemical step. In the extreme case, the chemical reaction may be so fast that all of R will be converted to Z, and no reverse peak will be observed. A classical example of such an EC mechanism is the oxidation of the drug chlorpromazine to form a radical cation that reacts with water to give an electroinactive sulfoxide. Ligand exchange reactions (e.g., of iron porphyrin complexes) occurring after electron transfer represent another example of such a mechanism.

Additional information on the rates of these (and other) coupled chemical reactions can be achieved by changing the scan rate (i.e. adjusting the experimental time scale). In particular, the scan rate controls the time spent between the switching potential and the peak potential (during which time the chemical reaction occurs). Hence, as illustrated in Figure 2.6, it is the ratio of the rate constant (of the chemical step) to the scan rate that controls the peak ratio. Most useful information is obtained when the reaction time lies within the experimental time scale. For scan rates between 0.02 and 200 V/s (common

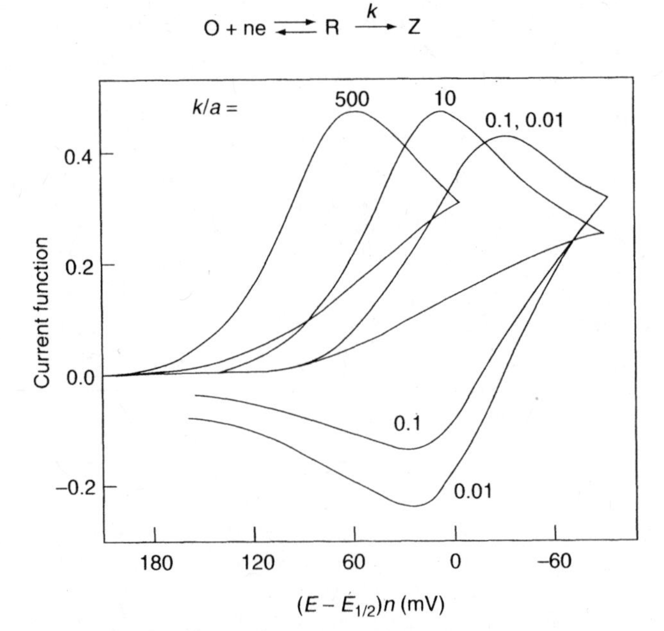

Figure 2.6 Cyclic voltammograms for a reversible electron transfer followed by an irreversible step for various ratios of chemical rate constant to scan rate k/a, where $a = nFv/RT$. (Reproduced with permission from Ref. 1.)

with conventional electrodes), the accessible time scale is around 0.1–1000 ms. Ultramicroelectrodes (discussed in Section 4.5.4) offer the use of much faster scan rates and hence the possibility of shifting the upper limit of follow-up rate constants measurable by cyclic voltammetry (3). For example, highly reactive species generated by the electron transfer, and alive for 25 ns, can be detected using a scan rate of 10^6 V/s. A wide variety of fast reactions (including isomerization and dimerization) can thus be probed. The extraction of such information commonly requires background subtraction to correct for the large charging-current contribution associated with ultrafast scan rates.

A special case of the EC mechanism is the catalytic regeneration of O during the chemical step:

$$\mathrm{O} + ne^- \rightleftharpoons \mathrm{R} \tag{2.8}$$

$$\mathrm{R} + \mathrm{A} \rightleftharpoons \mathrm{O} \tag{2.9}$$

An example of such a catalytic EC process is the oxidation of dopamine in the presence of ascorbic acid (4). The dopamine quinone formed in the redox step is reduced back to dopamine by the ascorbate ion. The peak ratio for such a catalytic reaction is always unity.

Other reaction mechanisms can be elucidated in a similar fashion. For example, for a CE mechanism, where a slow chemical reaction precedes the electron transfer, the ratio of $i_{p,r}/i_{p,f}$ is generally larger than one, and approaches unity as the scan rate decreases. The reverse peak is seldom affected by the coupled reaction, while the forward one is no longer proportional to the square root of the scan rate.

ECE processes, with a chemical step being interposed between electron transfer steps

$$O_1 + ne^- \rightleftharpoons R_1 \rightarrow O_2 + ne^- \rightarrow R_2 \quad (2.10)$$

are also easily explored by cyclic voltammetry, because the two redox couples can be observed separately. The rate constant of the chemical step can thus be estimated from the relative sizes of the two cyclic voltammetric peaks.

Many anodic oxidations involve an ECE pathway. For example, the neurotransmitter epinephrine can be oxidized to its quinone, which proceeds via cyclization to leucoadrenochrome. The latter can rapidly undergo electron transfer to form adrenochrome (5). The electrochemical oxidation of aniline is another classical example of an ECE pathway (6). The cation radical thus formed rapidly undergoes a dimerization reaction to yield an easily oxidized *p*-aminodiphenylamine product. Another example (of industrial relevance) is the reductive coupling of activated olefins to yield a radical anion, which reacts with the parent olefin to give a reducible dimer (7). If the chemical step is very fast (in comparison to the electron transfer process), the system behaves as an EE mechanism (of two successive charge transfer steps). Table 2.1 summarizes common electrochemical mechanisms involving coupled chemical reactions. Powerful cyclic voltammetric computational simulators, exploring the behavior of virtually any user-specific mechanism have been developed (9). Such simulated voltammograms can be compared with and fitted to the experimental ones. The new software also provides "movie"-like presentations of the corresponding continuous changes in the concentration profiles.

2.1.3 Study of Adsorption Processes

Cyclic voltammetry can also be used for evaluating the interfacial behavior of electroactive compounds. Both reactant and product can be involved in an adsorption–desorption process. Such interfacial behavior can occur in studies of numerous organic compounds, as well as of metal complexes (if the ligand is specifically adsorbed). For example, Figure 2.7 illustrates repetitive cyclic voltammograms, at the hanging mercury drop electrode, for riboflavin in a sodium hydroxide solution. A gradual increase of the cathodic and anodic peak currents is observed, indicating progressive adsorptive accumulation at the surface. Note also that the separation between the peak potentials is smaller than expected for solution-phase processes. Indeed, ideal Nernstian behavior of surface-confined nonreacting species is manifested by symmetric

TABLE 2.1 Electrochemical Mechanisms Involving Coupled Chemical Reactions

1. Reversible electron transfer, no chemical complications:

$$O + ne^- \rightleftharpoons R$$

2. Reversible electron transfer followed by a reversible chemical reaction—E_rC_r mechanism:

$$O + ne^- \rightleftharpoons R$$
$$R \underset{k_{-1}}{\overset{k_1}{\leftrightarrow}} Z$$

3. Reversible electron transfer followed by an irreversible chemical reaction—E_rC_i mechanism:

$$O + ne^- \rightleftharpoons R$$
$$R \overset{k}{\leftrightarrow} Z$$

4. Reversible chemical reaction preceding a reversible electron transfer—C_rE_r mechanism:

$$Z \underset{k_{-1}}{\overset{k_1}{\leftrightarrow}} O$$
$$O + ne^- \rightleftharpoons R$$

5. Reversible chemical reaction preceding an irreversible electron transfer—C_rE_i mechanism:

$$Z \underset{k_{-1}}{\overset{k_1}{\leftrightarrow}} O$$
$$O + ne^- \rightleftharpoons R$$

6. Reversible electron transfer followed by an irreversible regeneration of starting materials—catalytic mechanism:

$$O + ne^- \rightleftharpoons R$$
$$R + Z \overset{k}{\leftrightarrow} O$$

7. Irreversible electron transfer followed by an irreversible regeneration of starting material:

$$O + ne^- \rightleftharpoons R$$
$$R + Z \overset{k}{\leftrightarrow} O$$

8. Multiple electron transfer with intervening chemical reaction—ECE mechanism:

$$O + n_1e^- \rightleftharpoons R \quad R \rightleftharpoons Y$$
$$Y + n_2e^- \rightleftharpoons Z$$

Source: Adapted with permission from Ref. 8.

cyclic voltammetric peaks ($\Delta E_p = 0$), and a peak half-width of $90.6/n$ mV (Fig. 2.8). The peak current is directly proportional to the surface coverage (Γ) and potential scan rate:

$$i_p = \frac{n^2F^2\Gamma A v}{4RT} \tag{2.11}$$

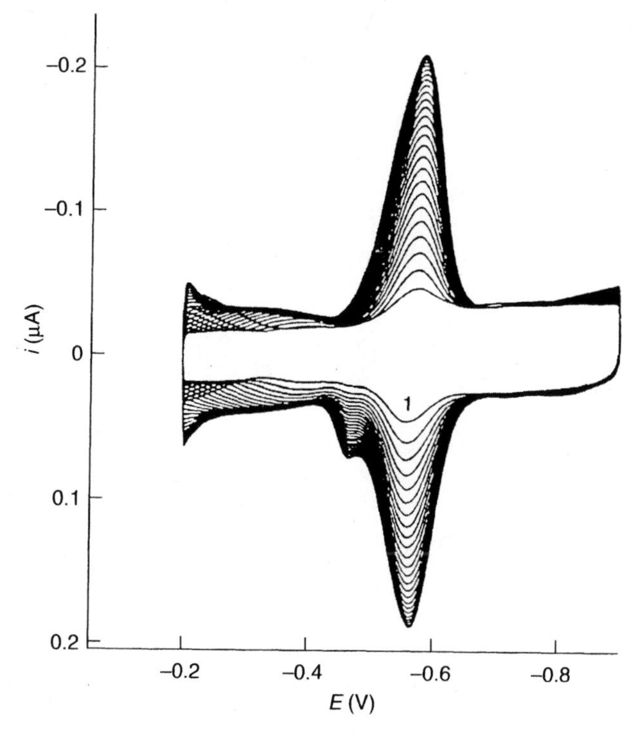

Figure 2.7 Repetitive cyclic voltammograms for 1×10^{-6} M riboflavin in a 1 mM sodium hydroxide solution. (Reproduced with permission from Ref. 10.)

Recall that a Nernstian behavior of diffusing species yields a $\nu^{1/2}$ dependence. In practice, the ideal behavior is approached for relatively slow scan rates, and for an adsorbed layer that shows no intermolecular interactions and fast electron transfers.

The peak area at saturation (i.e., the quantity of charge consumed during the reduction or adsorption of the adsorbed layer) can be used to calculate the surface coverage:

$$Q = nFA\Gamma \tag{2.12}$$

This can be used for calculating the area occupied by the adsorbed molecule and hence to predict its orientation on the surface. The surface coverage is commonly related to the bulk concentration via the adsorption isotherm. One of the most frequently used at present is the Langmuir isotherm

$$\Gamma = \Gamma_{\mathrm{m}} \frac{BC}{1 + BC} \tag{2.13}$$

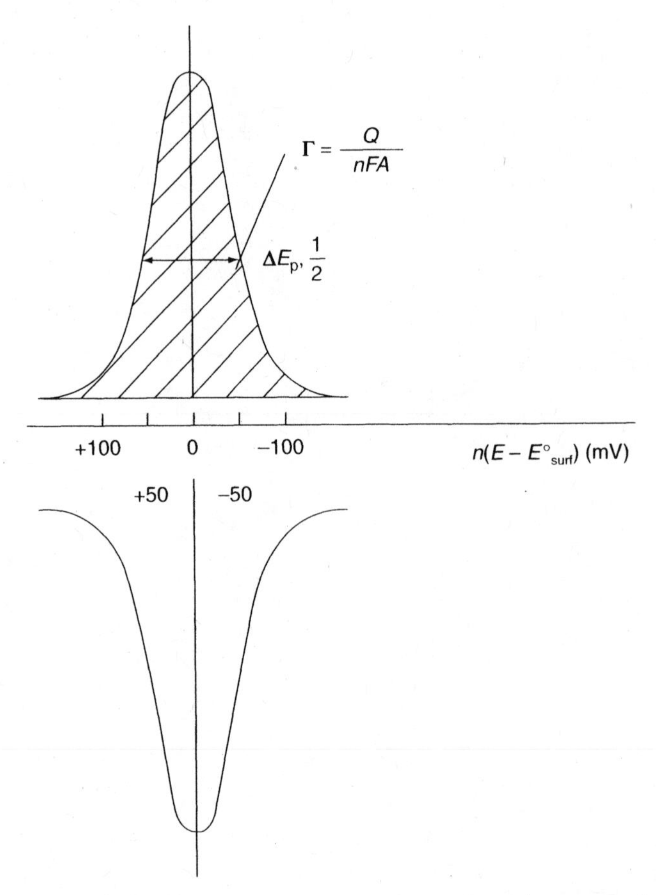

Figure 2.8 Ideal cyclic voltammetric behavior for a surface layer on an electrode. The surface coverage Γ can be obtained from the area under the peak. (Reproduced with permission from Ref. 11.)

where Γ_m is the surface concentration corresponding to a monolayer coverage (mol/cm^2) and B is the adsorption coefficient. A linearized isotherm, $\Gamma = \Gamma_m BC$, is obtained for low adsorbate concentrations (i.e., when $1 \gg BC$). The Langmuir isotherm is applicable to a monolayer coverage and assumes that there are no interactions between adsorbed species. Other isotherms (e.g., of Frumkin or Temkin) take into account such interactions. Indeed, the Langmuir isotherm is a special case of the Frumkin isotherm when no interactions exist. When either the reactant (O) or the product (R) (but not both) is adsorbed, one expects to observe a postpeak or prepeak, respectively (at potentials more negative or positive than the diffusion-controlled peak).

Equations have been derived for less ideal situations, involving quasi- and irreversible adsorbing electroactive molecules and different strengths of adsorption of the reactant and product (11–14). The rates of fast adsorption processes can be characterized by high-speed cyclic voltammetry at ultramicroelectrodes (15).

Two general models can describe the kinetics of adsorption. The first model involves fast adsorption with mass transport control, while the other involves kinetic control of the system. Under the latter (and Langmuirian) conditions, the surface coverage of the adsorbate at time t, Γ_t, is given by

$$\Gamma_t = \Gamma_e(1 - \exp(-k'C_t t)) \tag{2.14}$$

where Γ_e is the surface coverage and k' is the adsorption rate constant.

The behavior and performance of chemically modified electrodes based on surface-confined redox modifiers and conducting polymers (Chapter 4), can also be investigated by cyclic voltammetry, in a manner similar to that for adsorbed species. For example, Figure 2.9 illustrates the use of cyclic voltammetry for in situ probing of the growth of an electropolymerized film. Changes in the cyclic voltammetric response of a redox marker (e.g., ferrocyanide) are commonly employed for probing the blocking/barrier properties of insulating films (such as self-assembled monolayers).

2.1.4 Quantitative Applications

Cyclic voltammetry can be useful also for quantitative purposes, based on measurements of the peak current [Eq. (2.1)]. Such quantitative applications

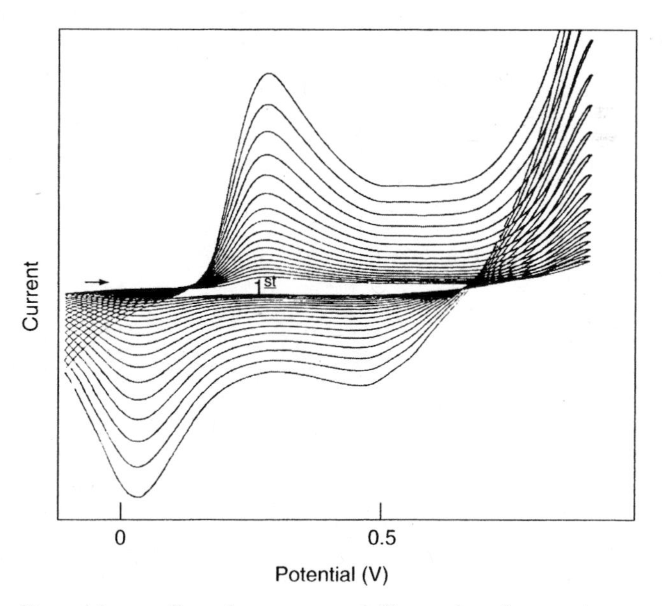

Figure 2.9 Repetitive cyclic voltammograms illustrating the continuous growth of polyaniline on a platinum surface.

require the establishment of the proper baseline. For neighboring peaks (of a mixture), the baseline for the second peak is obtained by extrapolating the current decay of the first one (in accordance to $t^{-1/2}$). Background reactions, primarily those associated with the double-layer charging and redox surface processes, limit the detection limit to around the 1×10^{-5} M level. Background-subtracted cyclic voltammetry can be employed for measuring lower concentrations (16). In particular, fast-scan (500–1000-V/s) background-subtracted cyclic voltammetry at carbon fiber microelectrodes is seeing increased use for the in vivo monitoring of neurotransmitters (such as dopamine or serotonin) in the human brain (17). Such coupling of digital background subtraction and fast voltammetric measurements provides the subsecond temporal resolution necessary for detecting dynamic concentration changes in the micromolar range occurring in the extracellular environment of the brain. The good temporal and chemical resolutions of such in vivo cyclic voltammetric experiments offer improved understanding of the chemistry of the brain. These repetitive scanning in vivo experiments generate large quantities of data, which are best represented as three-dimensional (potential, current, time) color contour images (18). For example, the temporal release of dopamine following an electrical stimulation is evidenced from the rapid increase in color around its peak potential. The ultrafast scanning also eliminates interferences from adsorption processes and chemical reactions that are coupled to the primary oxidation reaction of catecholamine neurotransmitters (19):

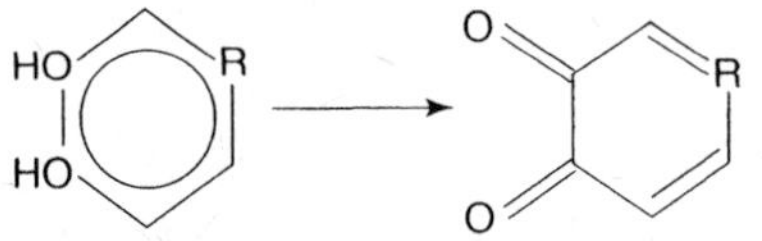

HO, HO, R (catechol) → O, O, R (o-quinone) $+ 2H^+ + 2e^-$ (2.15)

For more detailed information on the theory of cyclic voltammetry, and the interpretation of cyclic voltammograms, see Refs. 1, 7, 20, and 21.

2.2 SPECTROELECTROCHEMISTRY

The coupling of optical and electrochemical methods, spectroelectrochemistry, has been employed since the early 1980s to investigate a wide variety of inorganic, organic, and biological redox systems (22,23). Such a combination of electrochemical perturbations with the molecular specificity of optical monitoring successfully addresses the limited structural information available from the current response. It can be extremely useful for the elucidation of reaction mechanisms, and for the delineation of kinetic and thermodynamic parameters. A variety of informative optical methods have thus been coupled with electrochemical techniques. While the following sections will focus primarily on transmission absorption UV–vis (ultraviolet–visible) spectroscopic procedures, powerful spectroelectrochemical data can be obtained in reflectance experiments (in which the light beam is reflected from the electrode surface), using vibrational spectroscopic investigations, as well as from luminescence and scattering spectrochemical studies.

2.2.1 Experimental Arrangement

Optically transparent electrodes (OTEs), which enable light to be passed through their surface and the adjacent solution, are the key for performing transmission spectroelectrochemical experiments. One type of OTE consists of a metal (gold, silver, nickel) micromesh containing small (10–30-μm) holes, which couples good optical transmission (over 50%) with good electrical conductivity. Such a minigrid is usually sandwiched between two microscopic slides, which form a thin-layer cell (Fig. 2.10). The resulting chamber, containing the electroactive species in solution, contacts a larger container that holds the reference and auxiliary electrodes. The OTE is placed in the spectrophotometer so that the optical beam is passed directly through the transparent electrode and the solution. The working volume of the cell is only 30–50 μL, and complete electrolysis of the solute requires only 30–60 s. Alternately, the OTE may consist of a thin (100–5000-Å) film of a metal (e.g., gold, platinum) or a semiconductor (e.g., tin oxide), deposited on a transparent material such as quartz or glass substrate. The film thickness is often selected as a compromise between its electrical conductivity and optical transmission.

Improvements in cell design have been reported, including the use of fiber optics for the illumination and collection of light near electrode surfaces (24), the fabrication of long-pathlength OTEs via drilling of a small hole through a solid conducting material for sensitive optical monitoring of weakly absorbing species (25,26), and the incorporation of open porous materials (particularly reticulated vitreous carbon) within a thin-layer compartment (27).

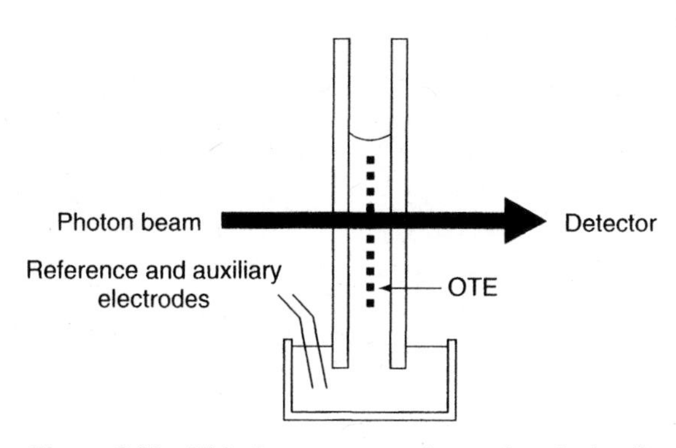

Figure 2.10 Thin-layer spectroelectrochemical cell.

2.2.2 Principles and Applications

The primary advantage of spectroelectrochemistry is the cross-correlation of information from the simultaneous electrochemical and optical measurements. In a typical experiment, one measures absorption changes resulting from species produced (or consumed) in the redox process. The change in absorbance is related to concentration and optical path length. Careful evaluation of the temporal absorbance response (A–t curve) during the electrochemical generation (or consumption) of an optically active species can yield extremely useful insight on reaction mechanisms and kinetics. Such experiments are particularly useful when the reactant and product have sufficiently different spectra.

Consider, for example, the general redox process:

$$\mathrm{O} + ne^- \rightleftharpoons \mathrm{R} \tag{2.16}$$

When the potential of the OTE is stepped to a value such that reaction (2.16) proceeds at a diffusion-controlled rate, the time-dependent absorbance of R is given by

$$A = \frac{2C_{\mathrm{O}}\varepsilon_{\mathrm{R}}D_{\mathrm{O}}^{1/2}t^{1/2}}{\pi^{1/2}} \tag{2.17}$$

where ε_R is the molar absorptivity of R and D_O and C_O are the diffusion coefficient and concentration of O, respectively. Hence, A increases linearly with the square root of time ($t^{1/2}$), reflecting the continuous generation of R at a rate determined by the diffusion of O to the surface. Equation (2.17) is valid when the generated species is stable. However, when R is a short-lived species (i.e., an EC mechanism), the absorbance response will be smaller than that expected from Eq. (2.17). The rate constant for its decomposition reaction can thus be calculated from the decrease in absorbance. Many other reaction mechanisms can be studied in a similar fashion from the deviation of the A–t curve from the shape predicted by Eq. (2.17). Such a potential-step experiment is known as *chronoabsorptometry*.

Thin-layer spectroelectrochemistry can be extremely useful for measuring the formal redox potential (E°) and n values. This is accomplished by spectrally determining the oxidized : reduced species concentration ratio ([O]/[R]) at each applied potential (from the absorbance ratio at the appropriate wavelengths). Since bulk electrolysis is achieved within a few seconds (under thin-layer conditions), the whole solution rapidly reaches an equilibrium with each applied potential (in accordance to the Nernst equation). For example, Figure 2.11 shows spectra for the complex $[Tc(dmpe)_2Br_2]^+$ in dimethylformamide using a series of potentials [dmpe is 1,2-bis(dimethylphosphine) ethane]. The logarithm of the resulting concentration ratio ([O]/[R]) can be plotted against the applied potential to yield a straight line, with an intercept corresponding

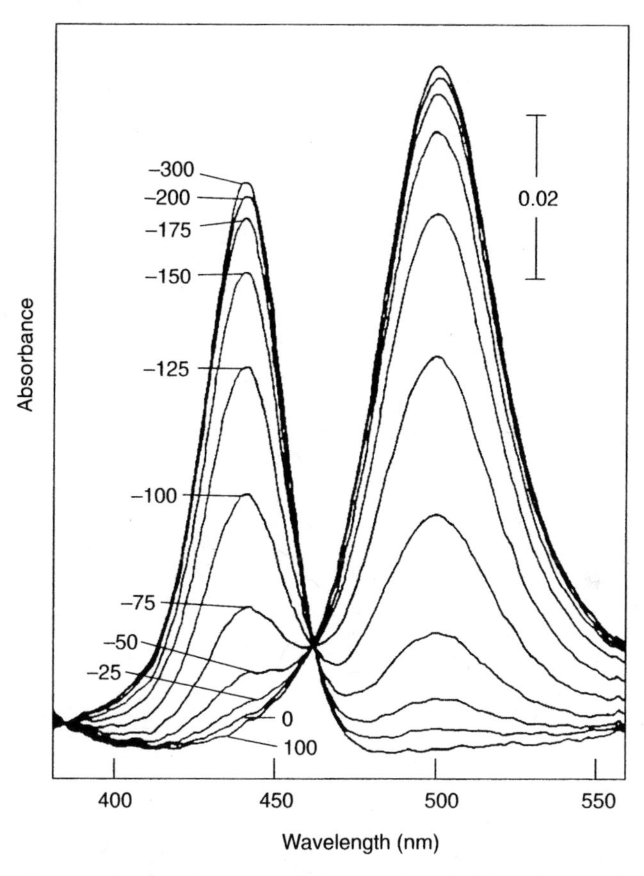

Figure 2.11 Spectra for a series of applied potentials (mV vs. Ag/AgCl) during thin-layer spectroelectrochemical experiment on 1.04 × 10^{-3}M $[Tc(III)(dmpe)_2Br_2]^+$. Medium: DMF containing 0.5 M TEAP. (Reproduced with permission from Ref. 28.)

to the formal potential. The slope of this Nernstian plot (0.059/n V) can be used to determine the n value.

Besides potential-step experiments, it is possible to employ linear potential scan perturbations of an OTE (28). This voltabsorptometric approach results in an optical analog of a voltammetric experiment. A dA/dE–E plot (obtained by differentiating the absorbance of the reaction product with respect to the changing potential) is morphologically identical to the voltammetric response for the redox process (Fig. 2.12). Depending on the molar absorptivity of the monitored species, the derivative optical response may afford a more sensitive tool than the voltammetric one. This concept is also not prone to charging-current background contributions and holds considerable promise for mechanism diagnosis and kinetic characterization of coupled chemical reactions.

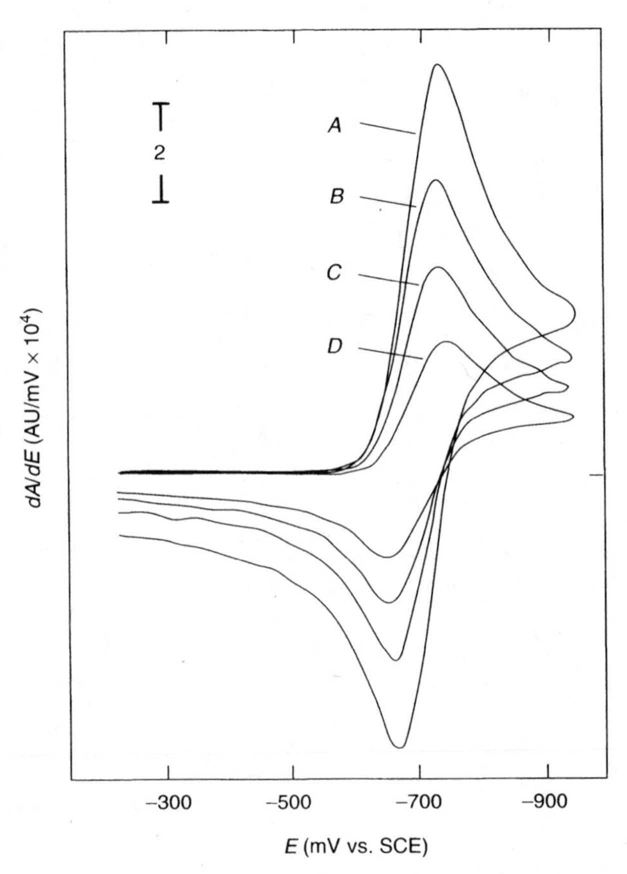

Figure 2.12 Plot of dA/dt versus E for 1.55×10^{-3} M methyl viologen at tin oxide OTE, using scan rates of 25 (A), 50 (B), 97.2 (C), and 265 (D) mV/s. (Reproduced with permission from Ref. 29.)

Spectroelectrochemical experiments can be used to probe various adsorption and desorption processes. In particular, changes in the absorbance accrue from such processes can be probed utilizing the large ratio of surface area to solution volume of OTEs with long optical pathlength (30). Additional information on such processes can be attained from the Raman spectroelectrochemical experiments described below.

In addition to transmission experiments, it is possible to use more sensitive reflectance protocols. In particular, in internal reflectance spectroscopy (IRS) the lightbeam is introduced to the electrode at an angle, and the spectrum are recorded from the reflected beam at the solid–solution interface. Prisms are used to let the radiation enter and leave. Besides its higher sensitivity, IRS is less prone to solution resistance effects.

Infrared spectroelectrochemical methods, particularly those based on Fourier transform infrared (FTIR), can provide structural information that

UV–vis absorbance techniques do not. FTIR spectroelectrochemicstry has thus been fruitful in the characterization of reactions occurring on electrode surfaces. The technique requires very thin cells to overcome solvent absorption problems.

Besides its widespread use for investigating the mechanism of redox processes, spectroelectrochemistry can be useful for analytical purposes. In particular, the simultaneous profiling of optical and electrochemical properties can enhance the overall selectivity of different sensing (31) and detection (32) applications. Such coupling of two modes of selectivity is facilitated by the judicious choice of the operating potential and wavelength.

2.2.3 Electrochemiluminescence

Electrochemiluminescence (ECL) is another useful technique for studying the fate of electrogenerated radicals that emit light. It involves the formation of light-emitting excited-state species as a result of highly and fast energetic electron transfer reactions of reactants formed electrochemically (33–36). Various organic and inorganic substances [e.g., polycyclic hydrocarbons, nitro compounds, luminol, $Ru(bpy)_3^{2+}$] can produce ECL on electron transfer from electrodes, in connection to the formation of radical ions. The electrogenerated radicals behave as very strong redox agents, and their reactions with each other or with other substances are sufficiently energetic to be able to populate excited states. ECL experiments are usually carried out by recording the spectra of the emitted light using a deoxygenated nonaqeous medium (e.g., highly purified acetonitrile or DMF). Operation in nonaqeous medium commonly involves the $Ru(bpy)_3^{2+}$ label because its ECL can be generated in this medium.

Analytical applications of ECL—relying on the linear dependence of the ECL light intensity and the reactant concentration—have also been realized (37). Since very low light levels can be measured (e.g., by single-photon counting methods), ECL offers extremely low detection limits. Such remarkable sensitivity has been exploited for a wide range of ECL-based immunoassays or DNA bioassays based on a $Ru(bpy)_3^{2+}$ label along with the tripropylamine (TPA) reagent (38). In order to generate light, $Ru(bpy)_3^{2+}$ and TPA are oxidized at the electrode surface to form a strong oxidant $Ru(bpy)_3^{3+}$ and a cation radical TPA^{+*}, respectively. The latter loses a proton and react with $Ru(bpy)_3^{3+}$ to form an excited state of $Ru(bpy)_3^{2+}$, which decays while releasing a photon at 620 nm (Fig. 2.13). The use of ECL as a detection method for liquid chromatography and microchip devices has also been documented (39,40).

In addition to UV–vis absorption measurements, other spectroscopic techniques can be used for monitoring the dynamics of electrochemical events or the fate of electrogenerated species. Particularly informative are the couplings of electrochemistry with electron spin resonance, nuclear magnetic resonance, and mass spectroscopy. A variety of specially designed cells have been con-

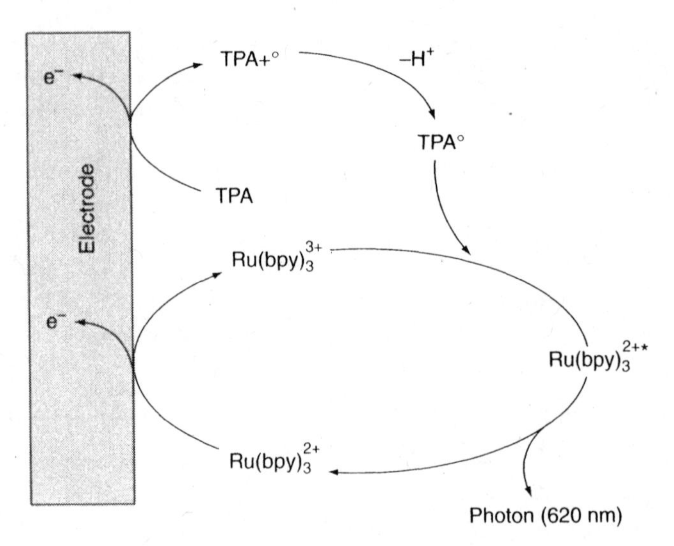

Figure 2.13 Electrochemiluminescence (ECL) reaction sequence, based on a $Ru(bpy)_3^{2+}$ label, commonly used for immunoassays or DNA bioassays.

structed to facilitate such studies, and several reviews have been published (41–45).

2.2.4 Optical Probing of Electrode–Solution Interfaces

Additional spectroscopic techniques can be used for probing the molecular structure of electrode–solution interfaces, as desired for understanding the fundamentals of electrode surfaces. The focus of these surface techniques is the correlation of the surface structure with electrochemical reactivity. Such surface-sensitive analytical tools can be classified as in situ or ex situ. In particular, the high sensitivity of molecular vibrations to the chemical environment has led to the widespread use of vibrational spectroscopies, such as surface-enhanced Raman scattering (SERS) for monitoring the surface composition before, during, and after the electrochemical experiment. In these experiments, a small fraction of the photons exchanges energy with the sample and are inelastically scattered, with a change of their wavelength characteristic of the energy change. Such Raman scattering effect can be enhanced by factors of $\leq 10^8$ when the compound is adsorbed on metal surfaces (46). The enhancement process is believed to result from the combination of several electromagnetic and chemical effects between the molecule and the metal surface. Since this scattering efficiency increases dramatically in the adsorbate state, Raman spectroelectrochemistry has been used primarily for investigating species adsorbed on electrodes (47). Another powerful in situ structural characterization technique, *X-ray adsorption fine structure* (EXAFS), refers to the modulation in the X-ray adsorption coefficient beyond the adsorption edge. Readers interested in these in situ techniques are referred to a 1991 monograph (48). Scanning electron microscopy (SEM) represents another widely used technique for obtaining ex situ information on the surface morphology and chemical composition (see, e.g., Fig. 4.20).

Other powerful ex situ techniques are based on the detection of charged particles derived from or interacting with the surface. Among these are low-energy electron diffraction (LEED), Auger electron spectroscopy (AES), and X-ray photoelectron spectroscopy (XPS), which are carried out in ultrahigh vacuum (UHV). In LEED, electrons directed at the sample at low energies (20–200 eV) diffract to produce a pattern unique to each substrate and adsorbed layer. In AES, an electron bombardment creates a vacancy in the electronic level close to the nucleus. This vacancy is filled by an electron coming from another electronic level, with the excess energy dissipated through ejection of a secondary electron (an Auger electron). The resulting energy spectrum consists of Auger peaks that are unique to each element. XPS [also known as *electron spectroscopy for chemical analysis* (ESCA)] can also provide atomic information about the surface region. In this technique, electrons are emitted from the sample on its irradiation with monochromatic X rays. The photon energy is related to the ionization (binding) energy E_B, the energy required to remove the electron from the initial state. Most elements (with the exception of hydrogen and helium) produce XPS signals with distinct E_B. In view of the limited penetration of X rays into solids, the technique gives useful information about surface structures or layers. The appearance of new XPS peaks can thus be extremely useful for studies of modified electrodes. The reliability of information gained by such ex situ analysis depends on a knowledge of what happens during vacuum exposure. Uncertainties associated with potential loss of material during such exposure have led to renew emphasis on direct (in situ) probes.

2.3 SCANNING PROBE MICROSCOPY

The more recently developed scanning probe microscopies (SPMs) appear to revolutionize the understanding of electrode processes. The purpose of this family of microscopes is to acquire high-resolution data of surface properties. The various scanning probe microscopies have similar subcomponents but different sensing probes. These high-resolution microscopies rely on sensing the interactions between a probe tip and the target surface, while scanning the tip across the surface. Different types of interactions can be sensed by the tip to yield different imaging signals. Such signals are displayed as gray scale portraits, reflecting the extent of the tip–surface interaction. With microcomputers, the image processing becomes possible in very short times. Among the various scanning probe microscopies, scanning tunneling microscopy, atomic force microscopy, and scanning electrochemical microscopy have been useful for imaging electrode surfaces directly (under potential control), and have thus dramatically improved the understanding of electrode reactions. Scanning probe microscopes have also been useful for creating nanostructures, through patterned movement and arrangement of nanoparticles and selected molecules.

2.3.1 Scanning Tunneling Microscopy

Scanning tunneling microscopy (STM) has revolutionized the field of surface science by allowing the direct imaging of surfaces on the atomic scale. The scanning tunneling microscope consists of a very sharp metallic tip that is moving over the surface of interest with a ceramic piezoelectric translator. The basis for its operation is the electron tunneling between the metal tip and the sample surface. The tunneling current (i_t) that flows when a voltage is applied between the sample and the tip is extremely sensitive to the sample–tip separation. In the simplest one-dimensional treatment i_t is given by

$$i_t \; \alpha \exp\left[(-4S\pi/h)(2m\phi)^{1/2}\right] \tag{2.18}$$

where S is the barrier width (equivalent to the shortest distance between the sample surface and the end of the tip), h is Planck's constant, m is the electron mass, and ϕ is the barrier height (equivalent to the local work function). In practice, i_t can change by a factor of ≥2 with a change of the tip–sample separation of only 1 Å. Accordingly, i_t tends to vary with the sample topography.

Although much of the early STM work has focused on investigating surfaces in vacuum, more recent work has demonstrated that surface images can also be obtained in liquid and air. In particular, the STM probing of electrode–electrolyte interfaces has attracted considerable attention (49–51). The ability of STM to offer structural information at the atomic level makes it highly suitable for in situ studies of time-dependent electrochemical processes, such as corrosion, electrodeposition, and adsorption, as well as surface modification, passivation, and activation. For example, Figure 2.14 shows a representative three-dimensional STM view of an electrochemically pretreated glassy carbon electrode, while Figure 2.15 illustrates an STM image of a spontaneously adsorbed alkanethiol monolayer on a gold surface. Useful insights into the structural–preparation relationships of conducting polymers can also be achieved by monitoring the growth of such films under different conditions (54). In addition to topographic information, the high sensitivity of the tunneling current to changes in the local work function (i.e., surface conductivity) offers a distinct visualization of composite electrode surfaces (55).

The more recent introduction of commercial STMs incorporating a potentiostat and an electrochemical cell has greatly facilitated in situ investigations of electrochemical processes. A block diagram of such STM–electrochemical system is shown in Figure 2.16. Coupled with powerful software, such instruments allow the simultaneous acquisition and display of the electrochemical

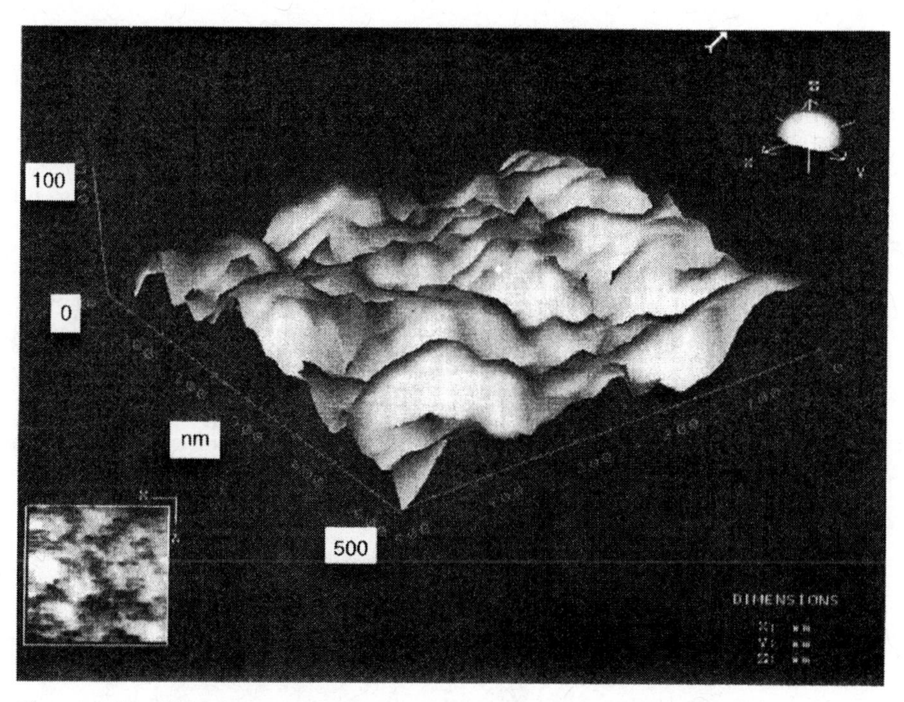

Figure 2.14 STM image of an electrochemically activated glassy carbon surface. (Reproduced with permission from Ref. 52.)

and topographic data. Extremely useful insights can thus be obtained by correlating the surface microstructures and the electrochemical reactivity. The interpretation of STM images requires extreme caution, and the tip should be shielded properly (from the electrolyte) to minimize the stray capacitance. Yet, the powerful coupling of STM and electrochemical systems offers many exciting future opportunities.

2.3.2 Atomic Force Microscopy

Atomic force microscopy (AFM) has become a standard technique to image with high resolution the topography of surfaces. It enables one to see nanoscopic surface features while the electrode is under potential control. This powerful probe microscopy operates by measuring the force between the probe and the samples (56,57). The probe consists of a sharp tip (made of silicon or silicon nitride) attached to a force-sensitive cantilever. The tip scans across the surface (by a piezoelectric scanner), and the cantilever deflects in response to force interactions between the tip and the substrate. Such deflection is monitored by bouncing a laser beam off it onto a photodetector. The measured force is attributed to repulsion generated by the overlap of the electron cloud at the probe tip with the electron cloud of surface atoms.

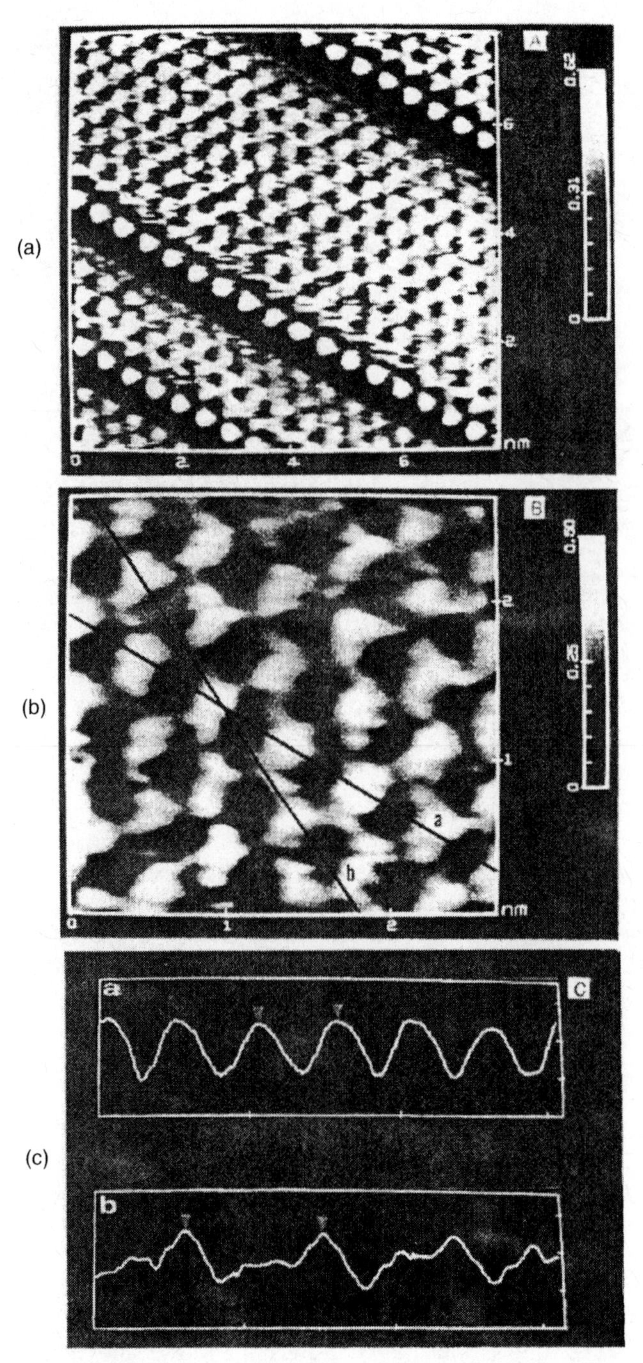

Figure 2.15 STM image of 7.7 × 7.7-nm (a) and 2.65 × 2.65-nm (b) sections of an ethanethiolate monolayer on a gold film; (c) contours of the image along the lines *a* and *b* in panel (b). (Reproduced with permission from Ref. 53.)

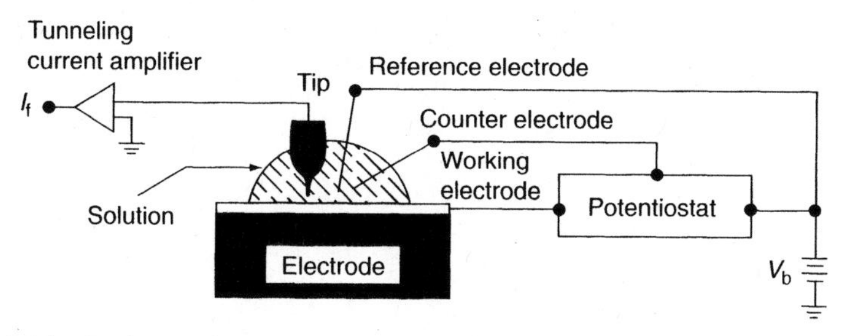

Figure 2.16 Design of a system for in situ electrochemical scanning tunneling microscopy.

It depends in part on the nature of the electrode, the distance between the electrode and the tip, any surface contamination, and the tip geometry. An image (revealing individual atoms) is created as the probe is translated across the surface, while sensing the interaction of the force fields by the cantilever beam. Such images can be formed by a constant force or height modes (with known or measured deflections of the cantilever, respectively). Since AFM does not involve passage of current between the tip and the surface, it is useful for exploring both insulating and conducting regions. The surface structure of a wide range of materials can thus be explored, irrespective of their conductivity. The technique has thus been extremely useful for observing changes in electrode surfaces caused by adsorption, etching, or underpotential deposition. Note, however, that in its conventional form, AFM lacks chemical specificity. The topographic imaging capability of AFM has been shown useful for monitoring changes in the height associated with ligand–receptor binding events (such as antibody–antigen recognition), indicating promise for label-free biochips (58). In addition to imaging applications, atomic force probes have found important applications ranging from dip-pen nanolithography (59) to microcantilever-based sensors (60). The former allows depositing "inks" (such as biomolecules) onto solid surfaces in high resolution with the AFM tip used as a "pen".

2.3.3 Scanning Electrochemical Microscopy

The scanning tunneling microscope (STM) has led to several other variants (61). Particularly attractive for electrochemical studies is scanning electrochemical microscopy (SECM) (62–65). In SECM, faradaic currents at an ultramicroelectrode tip are measured while the tip is moved (by a piezoelectric controller) in close proximity to the substrate surface that is immersed in a solution containing an electroactive species (Fig. 2.17). These tip currents are a function of the conductivity and chemical nature of the substrate, as well as of the tip–substrate distance. The images thus obtained offer valuable insights into the microdistribution of the electrochemical and chemical activity, as well

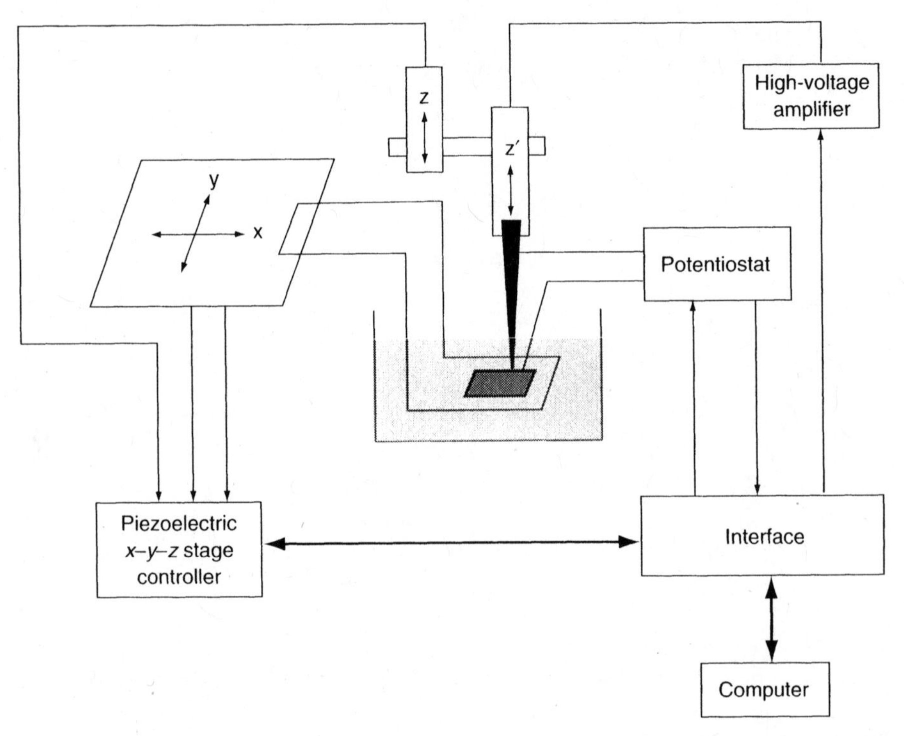

Figure 2.17 Design of a scanning electrochemical microscope. (Reproduced with permission from Ref. 66.)

as the substrate topography. A wide range of important applications involving different electrochemical systems have thus been developed.

The most common version of SECM, the feedback mode, involves recycling of an electroactive material between the tip and substrate surfaces (Fig. 2.18). When the microelectrode is distant from the surface by several electrode diameters, a steady-state current $i_{T,\infty}$ is observed at the tip (Fig 2.18a). When the tip is brought near a conductive substrate (held at sufficiently positive potential), the tip-generated product R can be oxidized back to O, and the tip current will be greater than $i_{T,\infty}$ (Fig. 2.18b). In contrast, when the tip passes over an insulating region (on the substrate), diffusion to the tip is hindered, and the feedback current diminishes (Fig. 2.18c). For example, Figure 2.19 displays a two-dimensional scan of a gold minigrid surface. The conducting gold lines are clearly observed from the enhanced recycling current. Alternately, in the collection mode, the tip is used only as a detector of species generated at the substrate. The distribution of the electrochemical activity of the surface can thus be mapped. The submicron resolution of SECM images is controlled by the size and shape of the tip, and it can be further improved by using digital image processing techniques. Yet, unlike STM or AFM, atomic resolution

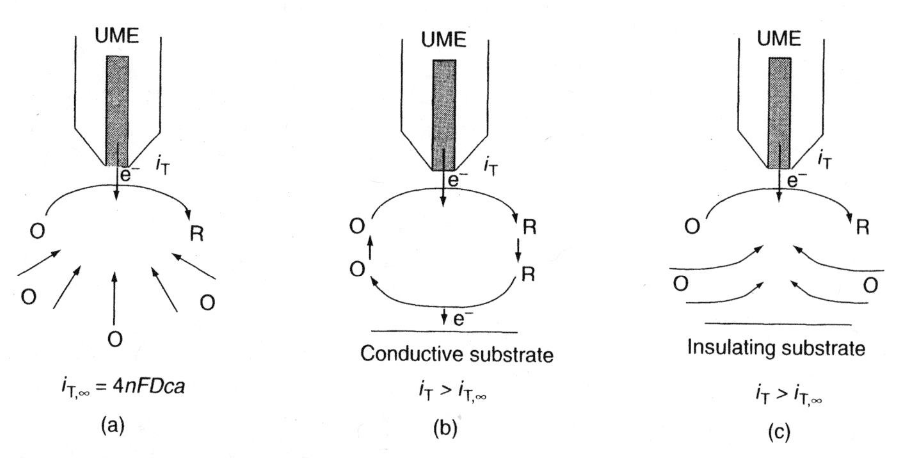

Figure 2.18 Principles of SECM: (a) tip far from the substrate surface; diffusion of O leads to steady-state current; (b) tip near a conductive substrate, with positive feedback of O; (c) tip near the insulating substrate; hindered diffusion of O. (Reproduced with permission from Ref. 65.)

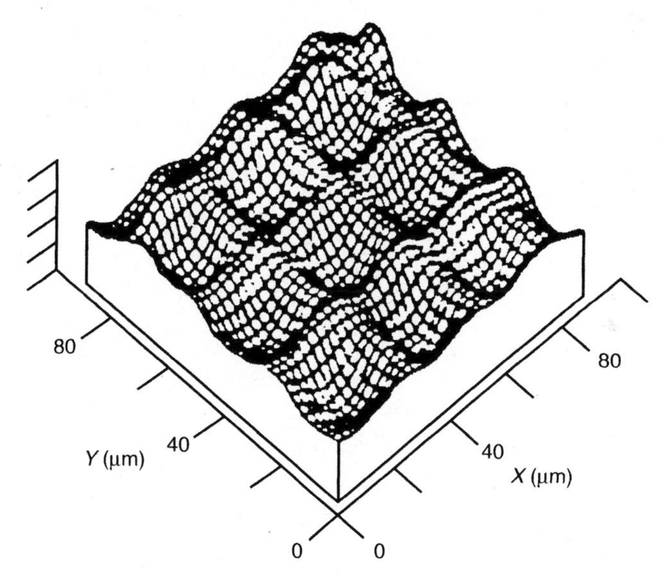

Figure 2.19 SECM image of a gold minigrid surface. (Reproduced with permission from Ref. 67.)

cannot be attained in SECM. Scanning electrochemical microscopy can also be used to investigate heterogeneous reaction kinetics. This is accomplished by forming a twin-electrode thin layer between the tip and a conducting substrate. Such configuration induces high rates of mass transfer and leads to tip currents limited by the intrinsic electron transfer rates. The volume reduction has been exploited also for electrochemical studies at the level of single molecules, which allow the elucidation of new effects that are not apparent in experiments involving a large number of molecules (68). For this purpose, the tip is insulated (e.g., with a wax) for trapping the single molecule in a tiny pocket (e.g., Fig. 2.20).

Scanning electrochemical microscopy can be applied also to study localized biological activity, for instance, as desired for in situ characterization of biosensors (69,70). For this purpose, the tip is used to probe the biological generation or consumption of electroactive species, such as the product of an enzymatic surface reaction. The technique holds promise for studying the effect of various chemical stimulations on cellular activities. SECM has been used for measuring localized transport through membranes, such as monitoring the ionic flux through the skin (71) or for imaging the respiratory activity single living cells, for instance, through measurements of oxygen at the sensing probe (72), or for visualizing DNA duplex spot regions of DNA microarrays (73).

The utility of potentiometric (pH-selective) tips has also been documented for imaging pH profiles, including those generated by enzymatic (urease) reactions (74). These and other (75) potentiometric tips are expected to probe different reactions that are not accessible with voltammetric tips, such as to determine local concentrations of electroinactive species. Unlike their voltammetric counterparts, potentiometric tips serve as purely passive sensors. Various electrochemical processes (e.g., electroplating, etching, corrosion) can also be characterized at high resolution while moving the tip over the substrate surface. In addition to its extensive use for surface characterization, SECM has been used as a microfabrication tool (76), with its tip acting as an electrochemical "pen" or "eraser." More recently introduced commercial

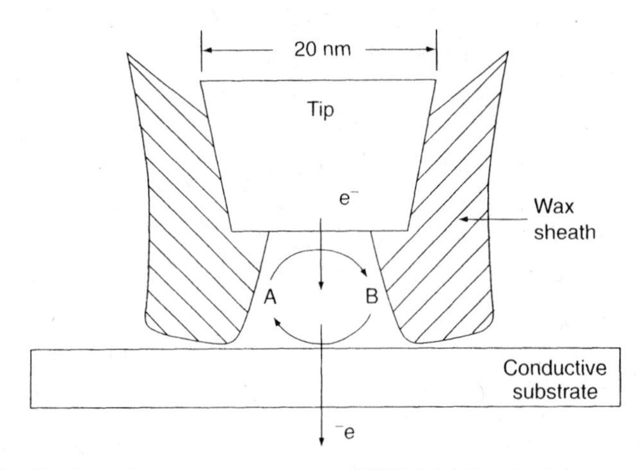

Figure 2.20 Single-molecule detection with SECM. Molecule A is trapped between the tip and the surface. (Reproduced with permission from Ref. 68.)

SECM instruments (77,78) will undoubtedly increase the scope and power of SECM. Further improvements in the power and scope of SECM has resulted from its coupling scanning probe or optical imaging techniques, such as AFM (57,79) or single-molecule fluorescence spectroscopy (80). The combined SECM-AFM technique offers simultaneous topographic and electrochemical imaging in connection to a probe containing a force sensor and an electrode component, respectively.

2.4 ELECTROCHEMICAL QUARTZ CRYSTAL MICROBALANCE

Electrochemical quartz crystal microbalance (EQCM) is a powerful tool for elucidating interfacial reactions based on the simultaneous measurement of electrochemical parameters and mass changes at electrode surfaces. The microbalance is based on a quartz crystal wafer, which is sandwiched between two electrodes, used to induce an electric field (Fig. 2.21). Such a field produces a mechanical oscillation in the bulk of the wafer. Surface reactions, involving minor mass changes, can cause perturbation of the resonant frequency of the crystal oscillator. The frequency change (Δf) relates to the mass change (Δm) according to the *Sauerbrey equation*:

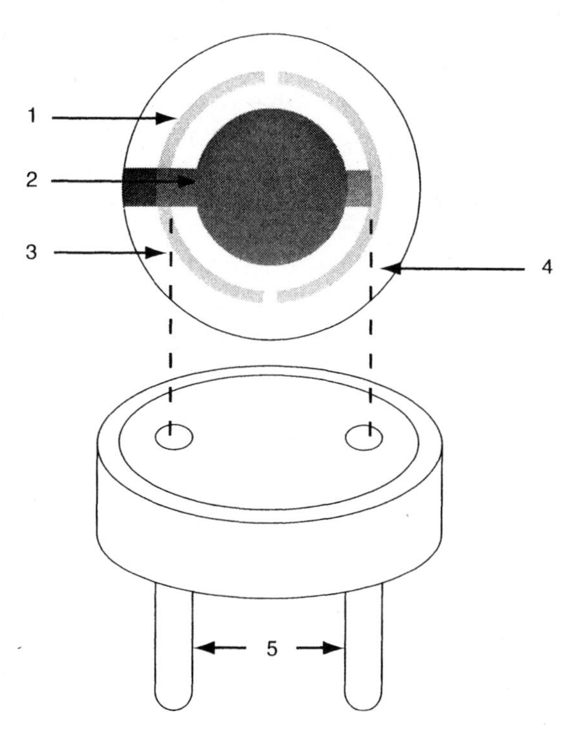

Figure 2.21 Quartz crystal microbalance: (1) the quartz crystal; (2) the gold electrode; (3,4) connecting metal wires; (5) the base.

$$\Delta f = -2\Delta mn f_0^2 / A\sqrt{\mu\rho} \tag{2.19}$$

where n is the overtone number, f_0 the base resonant frequency of the crystal (prior to the mass change), A is the area (cm^2), μ is the shear modulus of quartz ($2.95 \times 10^{11}\,g\,cm^{-1}\,s^{-1}$), and ρ is the density of quartz ($2.65\,g/cm^3$). As expected from the negative sign, decreases in mass correspond to increases in frequency and vice versa. The Sauebrey equation forms the basis for the excellent mass sensitivity of the EQCM. In situ mass changes of $1\,ng/cm^2$ can thus be detected. The EQCM is very useful for probing processes that occur uniformly across the surface. Numerous surface reactions have thus been investigated, including deposition or dissolution of surface layers and various uptake processes (such as doping/undoping of conducting polymers or ion exchange reactions at polymer films). Such changes can be probed using various controlled-potential or controlled-current experiments. In these experiments, one of the electrodes (on the wafer) contacts the solution and serves as the working electrode in the electrochemical cell, to allow simultaneous frequency and current measurements. For example, Figure 2.22 displays the frequency (mass) vs. potential profiles, and the corresponding cyclic voltammograms, during the uptake of a multiply charged complex ion at an ion exchanger coated electrode. Other useful examples of probing the uptake of mobile species by polymer-coated electrodes were given by Hillman et al. (82). Application of the Sauerbrey equation to the study of polymeric films in solutions requires adherence to the rigid-film approximation (i.e., behavior of elastic, solvent-free thin layer). Such approximation is valid when the thickness of the polymeric layer is small compared to the thickness of the crystal, and the measured frequency change is small with respect to the resonant frequency of the unloaded crystal. Mass changes of $\leq$0.05% of the crystal mass commonly meet this approximation. In the absence of molecular specificity EQCM cannot be used for molecule-level characterization of surfaces. Electrochemical quartz crystal microbalance devices also hold promise for the task of affinity-based chemical sensing, as they allow simultaneous measurements of both the mass and current. The principles and capabilities of EQCM have been reviewed (83,84). The combination of EQCM with scanning electrochemical microscopy has also been reported recently for studying the dissolution and etching of various thin films (85). The development of multichannel quartz crystal microbalance (86), based on arrays of resonators, should further enhance the scope and power of EQCM.

2.5 IMPEDANCE SPECTROSCOPY

Impedance spectroscopy is an effective technique for probing the features of chemically-modified electrodes and for understanding electrochemical reaction rates (87,88). Impedance is the totally complex resistance encountered

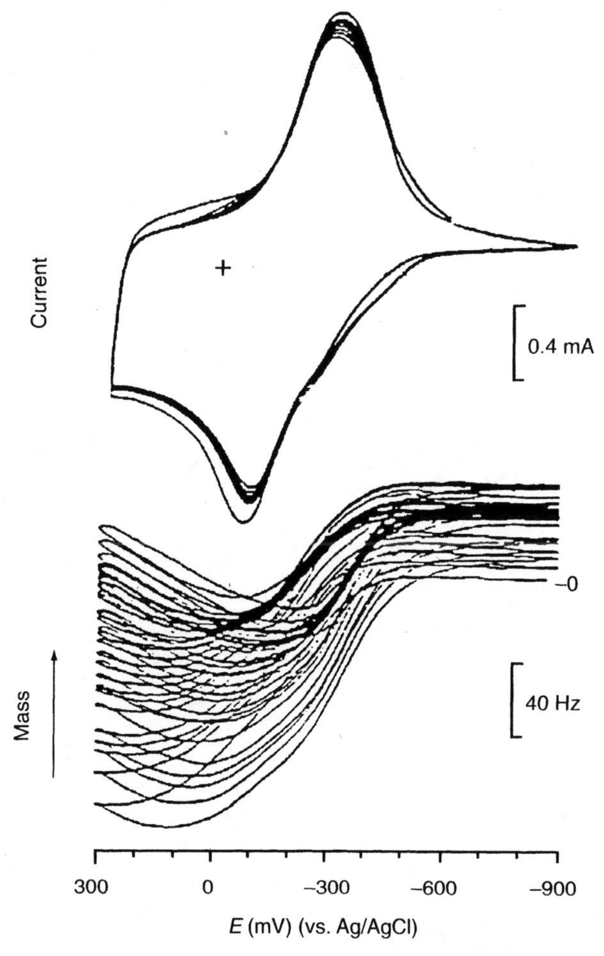

Figure 2.22 EQCM (bottom) and cyclic voltammetry (top) profiles at an ion-exchanger-coated electrode in the presence of 6×10^{-3} M $Ru(NH_3)_6Cl_6$. (Reproduced with permission from Ref. 81.)

when a current flows through a circuit made of combinations of resistors, capacitors, or inductors. Electrochemical transformations occurring at the electrode–solution interface can be modeled using components of the electronic equivalent circuitry that correspond to the experimental impedance spectra. Particularly useful to model interfacial phenomena is the Randles and Ershler electronic equivalent-circuit model (Fig. 2.23). This includes the double-layer capacitance C_d, the ohmic resistance of the electrolyte solution R_s, the electron transfer resistance R_p, and the Warburg impedance W resulting from the diffusion of ions from the bulk solution to the electrode surface. The impedance of the interface, derived by application of Ohm's law, consists of two parts, a real number Z' and an imaginary one, Z'':

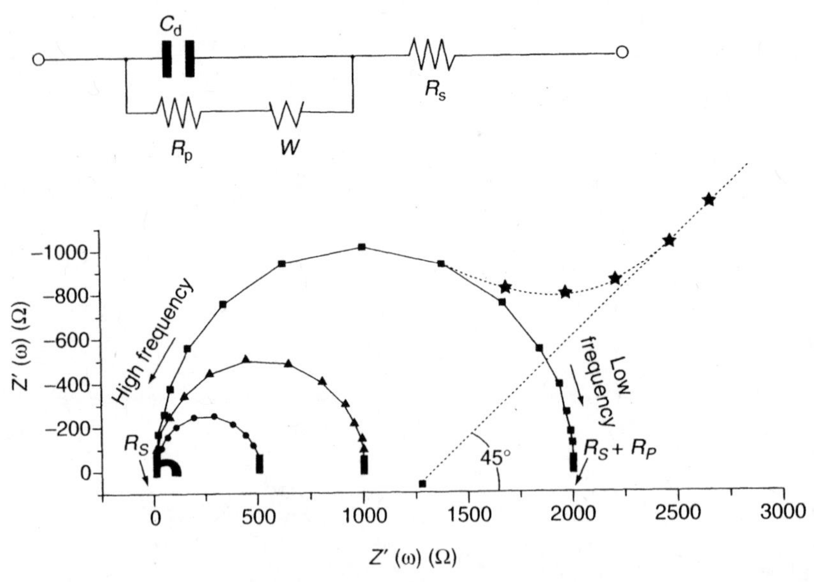

Figure 2.23 Faradaic impedance spectra presented in the form of Nyquist plots, along with the electronic equivalent circuit of the electrified interface. (Reproduced with permission from Ref. 87.)

$$Z(\omega) = R_s + R_p/(1+\omega^2 R_p^2 C_d^2) - j\omega R_p^2 C_d/(1+\omega^2 R_p^2 C_d^2) = Z' + jZ'' \qquad (2.20)$$

where $j = \sqrt{-1}$.

Impedance spectroscopy involves the application of a small-amplitude perturbing sinusoidal voltage signal (at a ω frequency) to the electrochemical cell and measuring the current response. The resulting faradaic impedance spectrum, known as a *Nyquist plot*, corresponds to the dependence of the imaginary number on the real number (e.g., Fig. 2.23), and contains extensive information about the electrified interface and the electron transfer reaction. Nyquist plots commonly include a semicircle region lying on the axis followed by a straight line. The semicircle portion (observed at higher frequencies) corresponds to the electron-transfer-limited process, while the straight line (characteristic of the low-frequency range) represents the diffusion-limited process. Such spectra can be used for extracting the electron transfer kinetics and diffusional characteristics. In the case of very fast electron transfer processes the impedance spectrum includes only the linear part, while very slow electron transfer processes are characterized by a large semicircular region. The diameter of the semicircle equals the electron transfer resistance. The intercepts of the semicircle with the Z_{re} axis correspond to those of R_s. In addition to fundamental electrochemical studies, the technique has been found extremely useful for transduction of bioaffinity events in connection to modern electrical immunosensors and DNA biosensors (88). Such transduction of bioaffinity events relies on the increased insulation of the electrode surface in respect to redox probes (e.g., ferrocyanide), present in the solution, on binding of large biomolecules (e.g., capture of an antigen that retards the electron transfer).

EXAMPLES

Example 2.1 The reversible oxidation of dopamine (DA) is a $2e^-$ process. A cyclic voltammetric anodic peak current of 2.2 μA is observed for a 0.4-mM solution of dopamine in phosphate buffer at a glassy carbon disk electrode of 2.6 mm^2 with a scan rate of 25 mV/s. What will i_p be for v = 100 mV/s and 1.2 mM DA?

Solution From Equation (2.1):

$$i_p = kCv^{1/2}$$
$$2.2 = k0.4(25)^{1/2}$$
$$k = 1.1$$

Under the new experimental conditions, i_p is given by

$$i_p = 1.1 \times 1.2 \times (100)^{1/2} = 13.2\,\mu A$$

Example 2.2 The following cyclic voltammogram was recorded for a reversible couple:

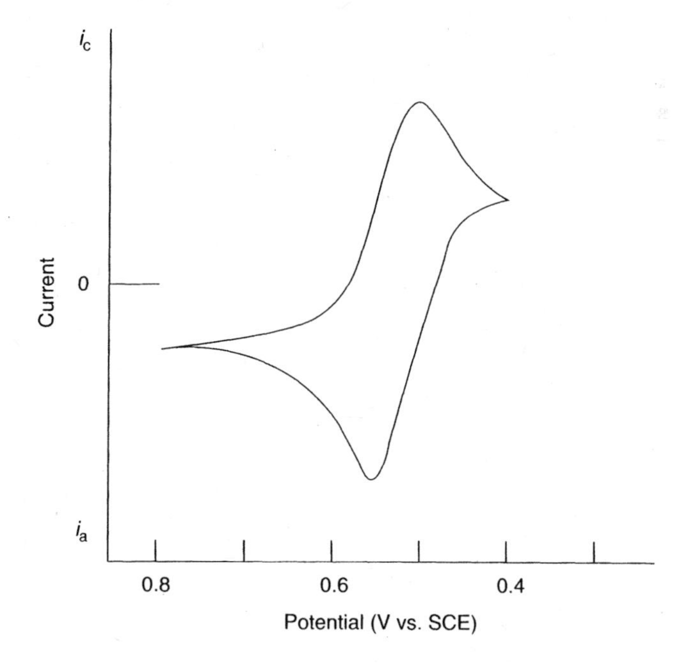

Calculate the number of electrons transferred and the formal potential for the couple.

Solution From Eq. (2.3):

$$n = 0.059/\Delta E_p = 0.059/(0.555 - 0.495) = 0.983 \cong 1.0$$

The formal potential can be calculated from Eq. (2.2):

$$E° = (0.555 + 0.495)/2 = 0.525\,\text{V}$$

Example 2.3 The electropolymeric growth of 2 ng polyphenol onto a gold QCM crystal (A = 1 cm^2; f_0 = 5 MHz) resulted in a frequency change of 12.5 Hz. Calculate the frequency change associated with the deposition of 4 ng polyphenol onto a 0.5-cm^2 crystal (f_0 = 8 MHz).

Solution From Eq. (2.19):

$$\Delta f = -K\Delta m f_0^2/A$$
$$12.5 = -K\,2 \times 5^2/1 \qquad K = -0.25$$

Under the new experimental conditions, Δf is given by

$$\Delta f = -0.25 \times 4 \times 8^2/0.5 = 128\,\text{Hz}$$

Example 2.4 A potential-step spectroelectrochemistry experiment using a reactant concentration of 2 mM generated a product with an absorbance (sampled after 25 s) of 0.8. Calculate the reactant concentration that yielded an absorbance of 0.4 on sampling at 16 s.

Solution From Equation (2.17):

$$A = KC_O t^{1/2} \qquad 0.8 = K\,2(25)^{1/2} \qquad K = 0.08$$

Accordingly

$$0.4 = 0.08 C_O (16)^{1/2} \qquad C_O = 5\,\text{mM}$$

Example 2.5 A thin-layer spectroelecrochemistry experiment for the $O + ne^- \rightarrow R$ process generated the following Nernstian plot:

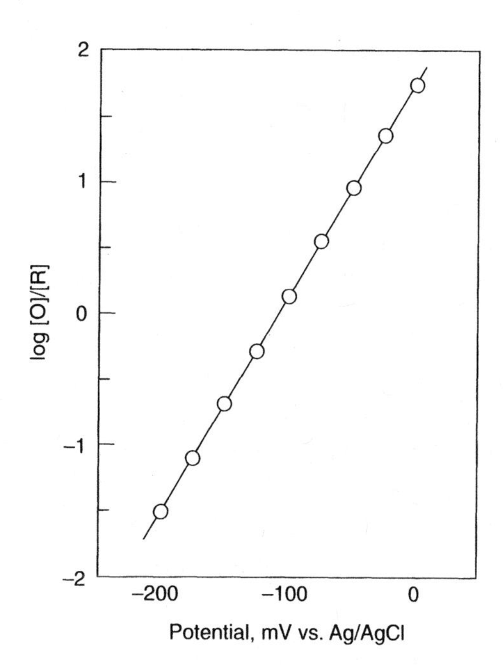

Calculate the number of electrons and formal potential of this redox process.

Solution The slope of this plot, 0.019 = 0.059/n, where $n = 3$. The intercept, 0.11 V, corresponds to the formal potential.

PROBLEMS

2.1 Explain and demonstrate how spectroelectrochemistry can provide useful information about a reaction mechanism involving a redox process followed by a chemical reaction (EC mechanism), involving decomposition of the reaction product. Draw absorbance–time signals for different rate constants of the decomposition reaction.

2.2 Which voltammetric technique can be used to estimate the surface coverage of an adsorbed molecule? How?

2.3 Draw an EQCM (mass–potential) profile for a metal deposition–stripping process during a cycling voltammetric scanning.

2.4 A cyclic voltammetric peak current of 12.5 μA was observed for the reversible reduction of a 1.5-mM lead solution using a 1.2-mm-diameter disk electrode and a 50 mV/s scan rate. Calculate the lead concentration that yields a peak current of 20.2 μA at 250 mV/s.

2.5 Discuss the difference between the feedback and generation/collection modes of SECM.

2.6 Explain how SECM images the microdistribution of the electrode activity of composite electrodes.

2.7 Describe how thin-layer spectroelectrochemistry is used to measure the values of E° and n.

2.8 Summarize the different features of cyclic voltammetric response for reversible and quasireversible systems.

2.9 Explain the use of cyclic voltammetry for estimating the values of E° and n for a reversible system.

2.10 What is the reason for the gradual increase of the cathodic and anodic cyclic voltammetric peak currents observed on repetitive scanning?

2.11 How would you use EQCM for elucidating the electrostatic desorption of anionic DNA molecules from gold electrodes?

2.12 Propose a SECM experiment for mapping the distribution of an oxidase enzyme within a carbon composite surface. (Note that the enzyme generates hydrogen peroxide in the presence of its substrate and oxygen.)

2.13 How does an increase in scan rate affect the ratio of peak currents (backward/forward) in a cyclic voltammetric experiment involving a redox process followed by a chemical reaction?

REFERENCES

1. Nicholson, R. S.; Shain, I., *Anal. Chem.* **36**, 706 (1964).
2. Echegoyen, L.; Echegoyen, L. E., *Acc. Chem. Res.* **31**, 593 (1998).
3. Andrieux, C. P.; Hapiot, P.; Savéant, J. M., *Electroanalysis* **2**, 183 (1990).
4. Gelbert, M. B.; Curran, D. J., *Anal. Chem.* **58**, 1028 (1986).
5. Hawley, M. D.; Tatawawdi, S. V.; Piekarski, S.; Adams, R. N., *J. Am. Chem. Soc.* **89**, 447 (1967).
6. Mohilner, D. M.; Adams, R. N.; Argersinger, W. R., *J. Am. Chem. Soc.* **84**, 3816 (1962).
7. Evans, D., *Acc. Chem. Res.* **10**, 313 (1977).
8. Mabbott, G., *J. Chem. Educ.* **60**, 697 (1983).
9. Rudolph, M.; Reddy, D.; Feldberg, S. W., *Anal. Chem.* **66**, 589A (1994).
10. Wang, J.; Luo, D. B.; Farias, P. A. M.; Mahmoud, J. S., *Anal. Chem.* **57**, 158 (1985).
11. Pearce, P. J.; Bard, A. J., *J. Electroanal. Chem.* **114**, 89 (1980).
12. Wopschall, R. H.; Shain, I., *Anal. Chem.* **39**, 1514 (1967).
13. Sluyters-Rehbach, M.; Sluyter, J. R., *J. Electroanal. Chem.* **65**, 831 (1975).
14. Brown, A. P.; Anson, F. C., *Anal. Chem.* **49**, 1589 (1977).
15. Stamford, J.; Hurst, P.; Kuhr, W.; Wightman, R. M., *J. Electroanal. Chem.* **265**, 291 (1989).
16. Baur, J.; Kristensen, E.; May, L.; Wiedemann, D.; Wightman, R. M., *Anal. Chem.* **60**, 1268 (1988).

17. Venton, B.; Wightman, R. M., *Anal. Chem.* **75**, 414A (2003).
18. Michael, D.; Travis, E.; Wightman, R. M., *Anal. Chem.* **70**, 586A (1998).
19. Jackson, B.; Dietz, S.; Wightman, R. M., *Anal. Chem.* **67**, 1115 (1995).
20. Heinze, J., *Angew Chem.* (*Int. Ed. Engl.*) **23**, 831 (1984).
21. Baldwin, R. P.; Ravichandran, K.; Johnson, R. K., *J. Chem. Ed.* **61**, 820 (1984).
22. Kuwana, T.; Heineman, W., *Acc. Chem. Res.* **9**, 241 (1976).
23. Heineman, W.; Hawkridge, F.; Blount, H., "Spectroelectrochemistry at optically transparent electrodes," in A. J. Bard, ed., *Electroanalytical Chemistry*, Marcel Dekker, New York, 1986, Vol. 13.
24. Van Dyke, D. A.; Cheng, H. Y., *Anal. Chem.* **60**, 1256 (1988).
25. Porter, M.; Kuwana, T., *Anal. Chem.* **56**, 529 (1984).
26. Brewster, J.; Anderson, J. L., *Anal. Chem.* **54**, 2560 (1982).
27. Norvell, V.; Mamantov, G., *Anal. Chem.* **49**, 1470 (1977).
28. Heineman, W. R., *J. Chem. Educ.* **60**, 305 (1983).
29. Bancroft, E.; Sidwell, J.; Blount, H., *Anal. Chem.* **53**, 1390 (1981).
30. Gui, Y. P.; Porter, M.; Kuwana, T., *Anal. Chem.* **57**, 1474 (1985).
31. Shi, Y.; Slaterbeck, A.; Seliskar, C.; Heineman, W. R., *Anal. Chem.* **69**, 3676 (1997).
32. Dewald, H. D.; Wang, J., *Anal. Chim. Acta* **166**, 163 (1984).
33. Faulkner, L.; Bard, A. J., "Electrochemiluminescence," in A. J. Bard, ed., *Electroanalytical Chemistry*, Marcel Dekker, New York, 1977, Vol. 10.
34. Velasco, J. G., *Electroanalysis* **3**, 261 (1991).
35. White, H. S.; Bard, A. J., *J. Am. Chem. Soc.* **104**, 6891 (1982).
36. Richter, M. M., *Chem. Rev.* **104**, 3003 (2004).
37. Knight, A. W.; Greenway, G., *Analyst* **119**, 879 (1994).
38. Kulmala, S.; Suomi, J., *Anal. Chim. Acta* **500**, 21 (2003).
39. Forry, S. P.; Wightman, R. M., *Anal. Chem.* **74**, 528 (2002).
40. Zhan, W.; Alvarez, J.; Crooks, R. M., *J. Am. Chem. Soc.* **124**, 13265 (2002).
41. Bagchi, R. N.; Bond, A. M.; Scholz, F., *Electroanalysis* **1**, 1 (1989).
42. Richards, J. A.; Evans, D. H., *Anal. Chem.* **47**, 964 (1965).
43. Chang, H.; Johnson, D. C.; Houk, R. S., *Trends Anal. Chem.* **8**, 328 (1989).
44. Regino, M. C.; Brajter-Toth, A., *Anal. Chem.* **69**, 5067 (1997).
45. Tong, Y.; Oldfield, E.; Wieckowski, A., *Anal. Chem.* **70**, 518A (1998).
46. Jeanmarie, D.; Van Duyne, R., *J. Electroanal. Chem.* **84**, 1 (1977).
47. McCreery, R. L.; Packard, R. T., *Anal. Chem.* **61**, 775A (1989).
48. Abruna, H. D., *Electrochemical Interfaces*, VCH Publishers, New York, 1991.
49. Arvia, A. J., *Surf. Sci.* **181**, 78 (1987).
50. Cataldi, T. R.; Blackham, I.; Briggs, A.; Pethica, J.; Hill, H. A., *J. Electroanal. Chem.* **290**, 1 (1990).
51. Wang, J., *Analyst* **117**, 1231 (1992).
52. Wang, J.; Martinez, T.; Yaniv, D.; McCormick, L. D., *J. Electroanal. Chem.* **278**, 379 (1990).
53. Widrig, C.; Alves, C.; Porter, M., *J. Am. Chem. Soc.* **113**, 2805 (1991).
54. Kim, T.; Yang, H.; Bard, A., *J. Electrochem. Soc.* **138**, L71 (1991).
55. Wang, J.; Martinez, T.; Yaniv, D.; McCormick, L. D., *J. Electroanal. Chem.* **286**, 265 (1990).

56. Hansma, H.; Weisenhorm, A.; Edmundson, A.; Gaub, H.; Hansma, P., *Clin. Chem.* **37**, 1497 (1991).
57. Gardner, C. E.; Macpherson, J. V., *Anal. Chem.* **74**, 576A (2002).
58. Jones, V.; Kenseth, J.; Porter, M. D.; Mosher, C. L.; Henderson, E., *Anal. Chem.* **70**, 1233 (1998).
59. Demers, L. M.; Ginger, D.; Park, S.; Li, Z.; Chung, S.; Mirkin, C. A., *Science* **296**, 1836 (2002).
60. Liu, W.; Montana, V.; Chapman, E. R.; Mohideen, U.; Parpura, V., *Proc. Natl. Acad. Sci. USA* **100**, 13621 (2003).
61. Wickramaasinghe, H., *Scientific Am.* **98** (Oct. 1989).
62. Engstrom, R.; Pharr, C., *Anal. Chem.* **61**, 1099A (1989).
63. Bard, A. J.; Denuault, G.; Lee, C.; Mandler, D.; Wipf, D., *Acc. Chem. Res.* **23**, 357 (1990).
64. Mirkin, M. V., *Anal. Chem.* **68**, 177A (1996).
65. Arca, M.; Bard, A. J.; Horrocks, B.; Richards, T.; Treichel, D., *Analyst* **119**, 719 (1994).
66. Liu, H.; Fan, F.; Lin, C.; Bard, A. J., *J. Am. Chem. Soc.* **108**, 3838 (1986).
67. Kwak, J.; Bard, A. J., *Anal. Chem.* **61**, 1794 (1989).
68. Fan, F.; Kwak, J.; Bard, A. J., *J. Am. Chem. Soc.* **118**, 9669 (1996).
69. Wang, J.; Wu, L.; Li, R., *J. Electroanal. Chem.* **272**, 285 (1989).
70. Pierce, D.; Unwin, P.; Bard, A. J., *Anal. Chem.* **64**, 1795 (1992).
71. Bath, B. D.; Scott, E. R.; Phipps, J. B.; White, J. S., *J. Pharm. Sci.* **89**, 1537 (2000).
72. Yasukawa, T.; Kaya, T.; Matsue, T., *Electroanalysis* **12**, 653 (2000).
73. Yamashita, K.; Takagi, M.; Uchida, K.; Kondo, H.; Takenak, S., *Analyst* **126**, 1210 (2001).
74. Horrocks, B.; Mirkin, M.; Pierce, D.; Bard, A. J.; Nagy, G.; Toth, K., *Anal. Chem.* **65**, 1304 (1993).
75. Wei, C.; Bard, A. J.; Nagy, G.; Toth, K., *Anal. Chem.* **67**, 1346 (1995).
76. Nowall, W.; Wipf, D.; Kuhr, W. G., *Anal. Chem.* **70**, 2601 (1998).
77. *Scanning Electrochemical Microscope*, CH Instruments, 1998.
78. Smith, J. P., *Anal. Chem.* **73**, 39A (2001).
79. Kranze, C.; Friedbacher, G.; Mizaikoff, B.; Lugstein, A.; Smoliner, J.; Bertagnolli, E., *Anal. Chem.* **73**, 2491 (2001).
80. Boldt, F.; Heinze, J.; Diez, M.; Peterson, J.; Börsch, M., *Anal. Chem.* **76**, 3473 (2004).
81. Basak, S.; Bose, C.; Rajeshwar, K., *Anal. Chem.* **64**, 1813 (1992).
82. Hillman, A. R.; Loveday, D.; Swann, M.; Bruckenstein, S.; Wildle, C. P., *Analyst* **117**, 1251 (1992).
83. Deakin, M.; Buttry, D., *Anal. Chem.* **61**, 1147A (1989).
84. Ward, M. D.; Buttry, D. A., *Science* **249**, 1000 (1990).
85. Cliffel, D.; Bard, A. J., *Anal. Chem.* **70**, 1993 (1998).
86. Tatsuma, T.; Watanabe, Y.; Oyama, N.; Kiakizaki, K.; Haba, M., *Anal. Chem.* **71**, 3632 (1999).
87. Park, S. M.; Yoo, J. S., *Anal. Chem.* **75**, 455A (2003).
88. Katz, E.; Willner, I., *Electroanalysis* **15**, 913 (2003).

3

CONTROLLED-POTENTIAL TECHNIQUES

The basis of all controlled-potential techniques is the measurement of the current response to an applied potential. A multitude of potential excitations (including a ramp, potential steps, pulse trains, a sine wave, and various combinations thereof) exists. The present chapter reviews those techniques that are widely used.

3.1 CHRONOAMPEROMETRY

Chronoamperometry involves stepping the potential of the working electrode from a value at which no faradaic reaction occurs to a potential at which the surface concentration of the electroactive species is effectively zero (Fig. 3.1a). A stationary working electrode and unstirred (quiescent) solution are used. The resulting current–time dependence is monitored. As mass transport under these conditions is solely by diffusion, the current–time curve reflects the change in the concentration gradient in the vicinity of the surface (recall Section 1.2). This involves a gradual expansion of the diffusion layer associated with the depletion of the reactant, and hence decreased slope of the concentration profile as time progresses (see Fig. 3.1b). Accordingly, the current (at a planar electrode) decays with time (Fig. 3.1c), as given by the *Cottrell equation*

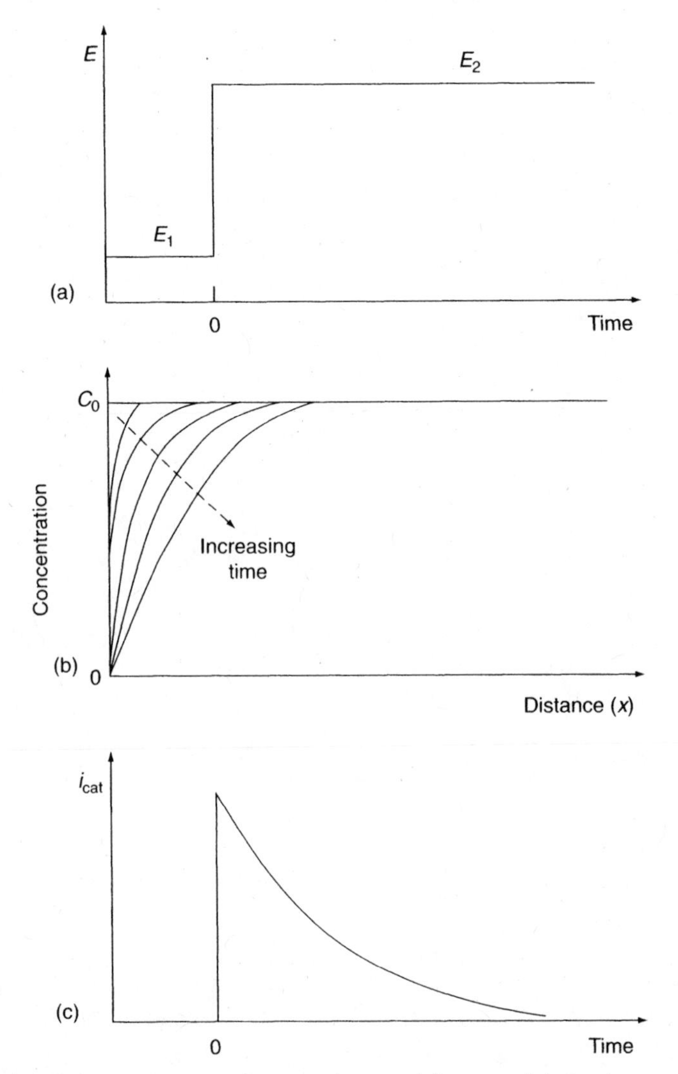

Figure 3.1 Chronoamperometric experiment: (a) potential–time waveform; (b) change in concentration profiles as time progresses; (c) the resulting current–time response.

$$i(t) = \frac{nFACD^{1/2}}{\pi^{1/2}t^{1/2}} = kt^{-1/2} \tag{3.1}$$

where n, F, A, C, D, and t are the number of electrons, Faraday's constant, the surface area, the concentration, the diffusion coefficient, and time, respectively. Such an $it^{1/2}$ constancy is often termed a "Cottrell behavior." Deviations from such behavior occur at long times (usually over 100s) as a result of natural convection effects, due to coupled chemical reactions, and when using non-planar electrodes or microelectrodes with high perimeter : area ratio (see Section 4.5.4). In the latter case, a time-independent current (proportional to the concentration) is obtained for $t > 0.1$ s, due to a large radial diffusion contribution. Similar considerations apply to spherical electrodes whose current response following potential step contains time-dependent and time-independent terms [Eq. (1.12)]. Recall also that for short values of t ($t < 50$ ms), the chronoamperometric signal contains an additional background contribution of the charging current [Eq. (1.49)]. This exponentially decaying charging current represents the main contribution to the response in the absence of an electroactive species.

Chronoamperometry is often used for measuring the diffusion coefficient of electroactive species or the surface area of the working electrode. Some analytical applications of chronoamperometry (e.g., in vivo bioanalysis) rely on pulsing of the potential of the working electrode repetitively at fixed time intervals. Some popular test strips for blood glucose (discussed in Chapter 6) involve potential-step measurements of an enzymatically liberated product (in connection with a preceding incubation reaction). Chronoamperometry can also be applied to the study of mechanisms of electrode processes. Particularly attractive for this task are reversal double-step chronoamperometric experiments (where the second step is used to probe the fate of a species generated in the first one).

The potential-step experiment can also be used to record the charge–time dependence. This is accomplished by integrating the current resulting from the potential step and adding corrections for the charge due to the double-layer charging (Q_{dl}) and reaction of the adsorbed species (Q_i):

$$Q = \frac{2nFACD^{1/2}t^{1/2}}{\pi^{1/2}} + Q_{dl} + Q_i \qquad (3.2)$$

Such a charge measurement procedure, known as *chronocoulometry*, is particularly useful for measuring the quantity of adsorbed reactants (because of the ability to separate the charges produced by the adsorbed and solution species). A plot of the charge (Q) versus $t^{1/2}$, known as an *Anson plot*, yields an intercept at $t = 0$ that corresponds to the sum of Q_{dl} and Q_i (Fig. 3.2). The former can be estimated by subtracting the intercept obtained in an identical experiment carried out in the blank solution.

3.2 POLAROGRAPHY

Polarography is a subclass of voltammetry in which the working electrode is dropping mercury. Because of the special properties of this electrode, particularly its renewable surface and wide cathodic potential range (see Chapters 3–5 for details), polarography has been widely used for the determination of

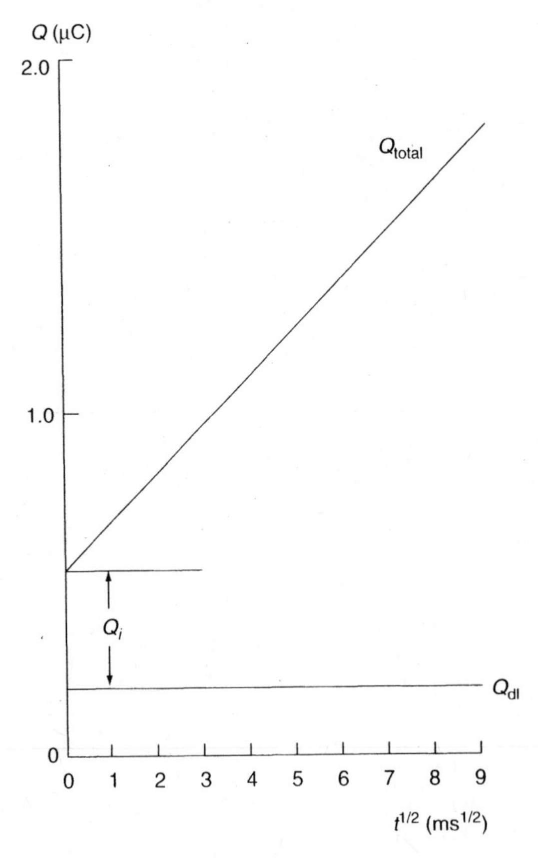

Figure 3.2 Chronocoulometric experiment: Anson plot of Q versus $t^{1/2}$.

many important reducible species. This classical technique was invented by J. Heyrovsky in Czechoslovakia in 1922, and had an enormous impact on the progress of electroanalysis (through many subsequent developments). Accordingly, Heyrovsky was awarded the 1959 Noble Prize in Chemistry.

The excitation signal used in conventional (DC) polarography is a linearly increasing potential ramp. For a reduction, the initial potential is selected to ensure that the reaction of interest does not take place. The potential is then scanned cathodically while the current is measured. Such current is proportional to the slope of the concentration–distance profile (see Section 1.2.1.2). At a sufficiently negative potential, reduction of the analyte commences, the concentration gradient increases, and the current rises rapidly to its limiting (diffusion-controlled) value. At this plateau, any analyte particle that arrives at the electrode surface instantaneously undergoes an electron transfer reaction, and the maximum rate of diffusion is achieved. The resulting polarographic wave is shown in Figure 3.3. The current oscillations reflect the growth and fall of the individual drops.

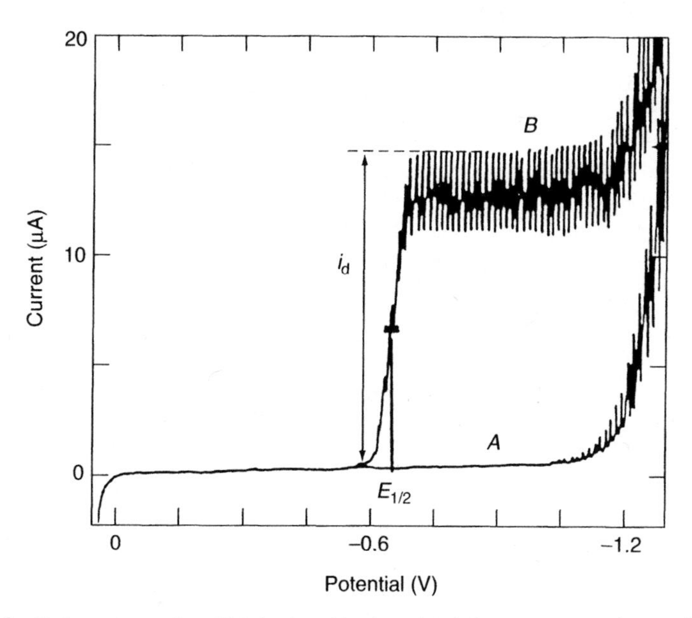

Figure 3.3 Polarograms for 1 M hydrochloric acid (*A*) and 4×10^{-4} M Cd^{2+} in 1 M hydrochloric acid (*B*); i_d represents the limiting current, while $E_{1/2}$ is the half-wave potential.

To derive the expression for the current response, one must account for the variation of the drop area with time

$$A = 4\pi\left(\frac{3mt}{4\pi d}\right)^{2/3} = 0.85(mt)^{2/3} \quad (3.3)$$

where t is the time and m and d are the mass flow rate and density of mercury, respectively. By substituting the surface area [from Eq. (3.3)] into the Cottrell equation [Eq. (3.1)], and replacing D by $7/3D$ (to account for the compression of the diffusion layer by the expanding drop), we can obtain the *Ilkovic equation* for the limiting diffusion current (1):

$$i_d = 708nD^{1/2}m^{2/3}t^{1/6}C \quad (3.4)$$

Here, i_d will have units of amperes (A), when D is in cm^2/s, m is in g/s, t is in seconds, and C is in mol/cm^3. This expression represents the current at the end of the drop life. The average current over the drop life is obtained by integrating the current of this time period:

$$i_{av} = 607nD^{1/2}m^{2/3}t^{1/6}C \quad (3.5)$$

To determine the diffusion current, it is necessary to subtract the residual current. This can be achieved by extrapolating the residual current prior to the wave or by recording the response of the deaerated supporting electrolyte (blank) solution. Standard addition or a calibration curve is often used for quantitation. Polarograms to be compared for this purpose must be recorded in the same way.

The potential where the current is one-half of its limiting value is called the *half-wave potential*, $E_{1/2}$. The half-wave potential (for electrochemically reversible couples) is related to the formal potential $E°$ of the electroactive species according to

$$E_{1/2} = E° + \frac{RT}{nF} \log(D_R / D_O)^{1/2} \tag{3.6}$$

where D_R and D_O are the diffusion coefficients of the reduced and oxidized forms of the electroactive species, respectively. Because of the similarity in the diffusion coefficients, the half-wave potential is usually similar to the formal potential. Thus, the half-wave potential, which is a characteristic of a particular species in a given supporting electrolyte solution, is independent of the concentration of that species. Therefore, by measuring the half-wave potential, one can identify the species responsible for an unknown polarographic wave. Typical half-wave potentials for several reducible organic functionalities, common in organic compounds, are given in Table 3.1. Compounds containing these functionalities are ideal candidates for polarographic measurements. (Additional oxidizable compounds can be measured using solid–electrode voltammetric protocols.) Since neutral compounds are involved, such organic polarographic reductions commonly involve hydrogen ions. Such reactions can be represented as

TABLE 3.1 Functional Groups Reducible at the DME

Class of Compounds	Functional	$E_{1/2}$ (V[a]) Group
Azo	—N=N—	–0.4
Carbon–carbon double bond[b]	—C=C—	–2.3
Carbon–carbon triple bond[b]	—C≡C—	–2.3
Carbonyl	>C=O	–2.2
Disulfide	S—S	–0.3
Nitro	NO_2	–0.9
Organic halides	C—X (X = Br, Cl, I)	–1.5
Quinone	C=O	–0.1

[a] Against the saturated calomel electrode at pH = 7.
[b] Conjugated with a similar bond or with an aromatic ring.

$$R + nH^+ + ne^- \rightleftharpoons RH_n \tag{3.7}$$

where R and RH_n are oxidized and reduced forms of the organic molecule. For such processes, the half-wave potential will be a function of pH (with a negative shift of about 59 mV/*n* for each unit increase in pH, due to decreasing availability of protons). Thus, in organic polarography, good buffering is vital for generating reproducible results. Reactions of organic compounds are also often slower and more complex than those for inorganic cations.

For the reduction of metal complexes, the half-wave potential is shifted to more negative potentials, reflecting the additional energy required for the complex decomposition. Consider the reversible reduction of a hypothetical metal complex, ML_p:

$$ML_p + ne^- + Hg \rightleftharpoons M(Hg) + pL \tag{3.8}$$

where L is the free ligand and p is the stoichiometric number. (The charges are omitted for simplicity.) The difference between the half-wave potential for the complexed and uncomplexed metal ion is given by (2)

$$(E_{1/2})_c - (E_{1/2})_{free} = \frac{RT}{nF} \ln K_d - \frac{RT}{nF} p \ln[L] + \frac{RT}{nF} \ln\left(\frac{D_{free}}{D_c}\right)^{1/2} \tag{3.9}$$

where K_d is the formation constant. The stoichiometric number can thus be computed from the slope of a plot $(E_{1/2})_c$ versus ln [L]. It is possible to exploit Eq. (3.9) to improve the resolution between two neighboring polarographic waves, based on a careful choice of the ligand and its concentration.

For reversible systems (with fast electron transfer kinetics), the shape of the polarographic wave can be described by the Heyrovsky–Ilkovic equation:

$$E = E_{1/2} + \frac{RT}{nF} \ln\left(\frac{i_d - i}{i}\right) \tag{3.10}$$

It follows from this equation that a plot of E versus log $[(i_d - i)/i]$ should yield a straight line with a slope of $0.059/n$ at 25°C. Such a plot offers a convenient method for the determination of n. In addition, the intercept of this line will be the half-wave potential. Another way to estimate n is to measure $(E_{3/4} - E_{1/4})$, which corresponds to $56.4/n$ mV for a reversible system ($E_{3/4}$ and $E_{1/4}$ are the potentials for which $i = 0.75 i_d$ and $i = 0.25 i_d$, respectively). It should be emphasized that many polarographic processes, especially those of organic compounds, are not reversible. For those that depart from reversibility, the wave is "drawn out," with the current not rising steeply, as is shown in Figure 3.3. The shape of the polarographic response for an irreversible reduction process is given by

$$E = E^{\circ} + \frac{RT}{\alpha nF} \ln\left[1.35k_{\mathrm{f}}\left(\frac{i_{\mathrm{d}} - i}{i}\right)\left(\frac{t}{D}\right)^{1/2}\right] \tag{3.11}$$

where α is the transfer coefficient and k_{f} is the rate constant of the forward reaction.

In a few instances, the polarographic wave is accompanied by a large peak (where the current rises to a maximum before returning to the expected diffusion current plateau). Such an undesired peak, known as the *polarographic maximum*, is attributed to a hydrodynamic flow of the solution around the expanding mercury drop, can be suppressed by adding a small amount of a surface-active material (such as Triton X-100).

When the sample solution contains more than one reducible species, diffusion currents resulting from each of them are observed. The heights of the successive waves can be used to measure the individual analytes, provided there is a reasonable difference (>0.2 V) between the half-wave potentials. The baseline for measuring the limiting current of the second species is obtained by extrapolation of the limiting current of the first process. With a potential window of ~2 V, five to seven individual polarographic waves could be observed. Solution parameters, such as the pH or concentration of complexing agents, can be manipulated to deliberately shift the peak potential and hence to improve the resolution of two successive waves. Successive waves are also observed for samples containing a single analyte that undergoes reduction in two or more steps (e.g., 1,4-benzodiazepine, tetracycline).

The background (residual) current that flows in the absence of the electroactive species of interest is composed of contributions due to double-layer charging process and redox reactions of impurities, as well as of the solvent, electrolyte, or electrode. The latter processes (e.g., hydrogen evolution and mercury oxidation) are those that limit the working potential range. In acidic solutions, the negative background limit shifts by approximately 59 mV per each pH unit to more positive potentials with decreasing pH. Within the working potential window, the charging current is the major component of the background (which limits the detection limit). It is the current required to charge the electrode–solution interface (which acts as a capacitor) on changing the potential or the electrode area (see Section 1.3). Thus, the charging current is present in all conventional polarographic experiments, regardless of the purity of reagents. Because of the negligible potential change during the drop life, the charging associated with the potential scan can be ignored. The value of the polarographic charging current thus depends on the time change of the electrode area:

$$i_{\mathrm{c}} = \frac{dq}{dt} = (E - E_{\mathrm{pzc}})C_{\mathrm{dl}}\frac{dA}{dt} \tag{3.12}$$

By substituting the derivative of the area with time [from Eq. (3.2)], one obtains

$$i_c = 0.00567(E - E_{pzc})C_{dl}m^{2/3}t^{-1/3} \quad (3.13)$$

Hence, the charging current decreases during the drop life, while the diffusion current increases (Fig. 3.4):

$$i_{total}(t) = i_d(t) + i_c(t) = kt^{1/6} + k't^{-1/3} \quad (3.14)$$

The analytical significance of the charging current depends on how large it is relative to the diffusion current of interest. When the analyte concentration is in the 10^{-4}–10^{-2} M range, the current is mostly faradaic, and a well-defined polarographic wave is obtained. However, at low concentrations of the analyte, the charging current contribution becomes comparable to the analytical signal, and the measurement becomes impossible. The charging current thus limits the detection limit of classical polarography to the 5×10^{-6}–1×10^{-5} M region. Lower detection limits are obtained for analytes with redox potentials closer to E_{pzc} [when i_c approaches its smaller value, Eq. (3.12)]. Advanced (pulse) polarographic techniques, discussed in Section 3.3, offer lower detection limits by taking advantage of the different time dependences of the analytical and charging currents [Eq. (3.14)]. Such developments have led to a decrease in the utility of DC polarography.

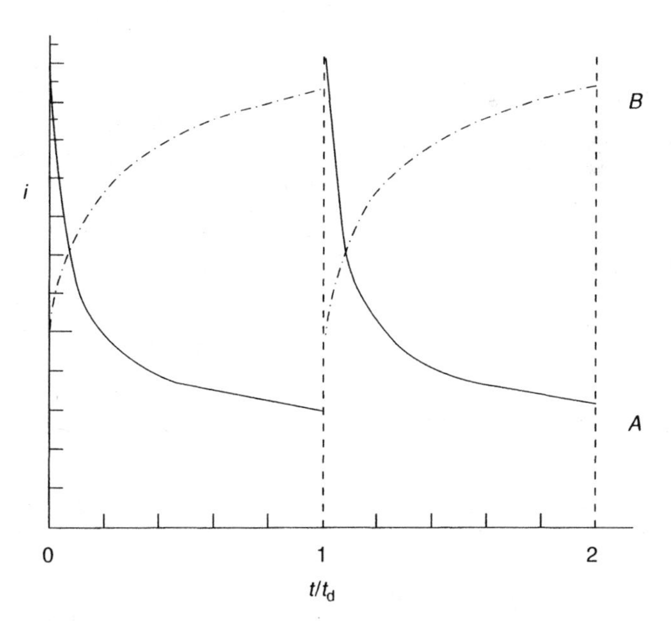

Figure 3.4 Variation of the charging and diffusion currents (A and B, respectively) during the lifetime of a drop.

3.3 PULSE VOLTAMMETRY

Pulse voltammetric techniques, introduced by Barker and Jenkin (3), are aimed at lowering the detection limits of voltammetric measurements. By substantially increasing the ratio between the faradaic and nonfaradaic currents, such techniques permit convenient quantitation down to the 10^{-8}M concentration level. Because of their greatly improved performance, modern pulse techniques have largely supplanted classical polarography in the analytical laboratory. The various pulse techniques are all based on a sampled current/potential-step (chronoamperometric) experiment. A sequence of such potential steps, each with a duration of about 50ms, is applied onto the working electrode. After the potential is stepped, the charging current decays rapidly (exponentially) to a negligible value, while the faradaic current decays more slowly. Thus, by sampling the current late in the pulse life, an effective discrimination against the charging current is achieved.

The difference between the various pulse voltammetric techniques is the excitation waveform and the current sampling regime. With both normal-pulse and differential-pulse voltammetry, one potential pulse is applied for each drop of mercury when the DME is used. (Both techniques can also be used at solid electrodes.) By controlling the drop time (with a mechanical knocker), the pulse is synchronized with the maximum growth of the mercury drop. At this point, near the end of the drop lifetime, the faradaic current reaches its maximum value, while the contribution of the charging current is minimal (based on the time dependence of the components).

3.3.1 Normal-Pulse Voltammetry

Normal-pulse voltammetry consists of a series of pulses of increasing amplitude applied to successive drops at a preselected time near the end of each drop lifetime (4). Such a normal pulse train is shown in Figure 3.5. Between

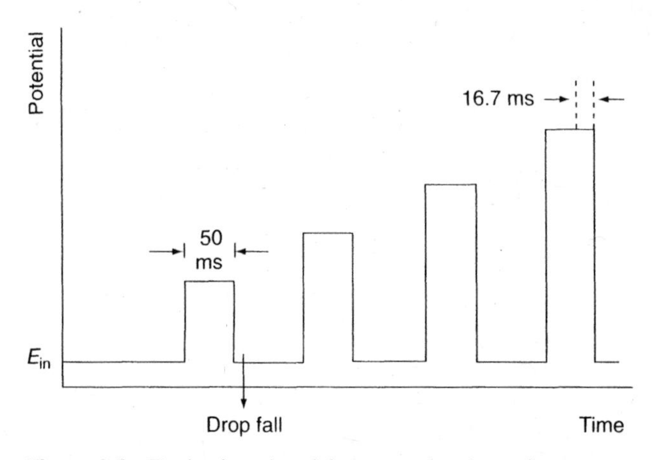

Figure 3.5 Excitation signal for normal-pulse voltammetry.

the pulses, the electrode is kept at a constant (base) potential at which no reaction of the analyte occurs. The amplitude of the pulse increases linearly with each drop. The current is measured about 40 ms after the pulse is applied, at which time the contribution of the charging current is nearly zero. In addition, because of the short pulse duration, the diffusion layer is thinner than that of DC polarography (i.e., greater flux of analyte) and hence the faradaic current is increased. The resulting voltammogram has a sigmoidal shape, with a limiting current given by a modified Cottrell equation:

$$i_{\mathrm{l}} = \frac{nFAD^{1/2}C}{\sqrt{\pi t_{\mathrm{m}}}} \tag{3.15}$$

where t_{m} is the time after application of the pulse where the current is sampled. This current can be compared to that measured in DC polarography:

$$\frac{i_{\mathrm{l,NP}}}{i_{\mathrm{l,dc}}} = \left(\frac{3t_{\mathrm{d}}}{7t_{\mathrm{m}}}\right)^{1/2} \tag{3.16}$$

This ratio predicts that normal-pulse polarography will be 5–10 times more sensitive than DC polarography (for typical values of t_{d} and t_{m}). Normal-pulse polarography may be advantageous also when using solid electrodes. In particular, by maintaining a low initial potential during most of the operation, it is possible to alleviate surface fouling problems (due to adsorbed reaction products).

A related technique, reverse-pulse voltammetry, has a pulse sequence that is a mirror image of that of normal-pulse voltammetry (5). In this case, the initial potential is on the plateau of the wave (i.e., where reduction occurs), and a series of positive-going pulses of decreasing amplitude is applied.

3.3.2 Differential-Pulse Voltammetry

Differential-pulse voltammetry is an extremely useful technique for measuring trace levels of organic and inorganic species. In differential-pulse voltammetry, fixed magnitude pulses—superimposed on a linear potential ramp—are applied to the working electrode at a time just before the end of the drop (Fig. 3.6). The current is sampled twice, just before the pulse application (at 1) and again late in the pulse life (after ~40 ms, at 2, when the charging current has decayed). The first current is instrumentally subtracted from the second, and this current difference [$\Delta i = i(t_2) - i(t_1)$] is plotted against the applied potential. The resulting differential-pulse voltammogram consists of current peaks, the height of which is directly proportional to the concentration of the corresponding analytes:

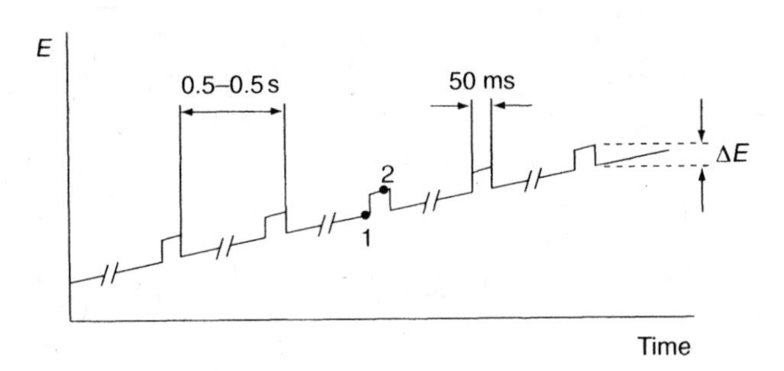

Figure 3.6 Excitation signal for differential-pulse voltammetry.

$$i_p = \frac{nFAD^{1/2}C}{\sqrt{\pi t_m}}\left(\frac{1-\sigma}{1+\sigma}\right) \quad (3.17)$$

where $\sigma = \exp[(nf/RT)(\Delta E/2)]$ (ΔE is the pulse amplitude). The maximum value of the quotient $(1 - \sigma)/(1+ \sigma)$, obtained for large pulse amplitudes, is unity (6).

The peak potential (E_p) can be used to identify the species, as it occurs near the polarographic half-wave potential:

$$E_p = E_{1/2} - \Delta E/2 \quad (3.18)$$

The differential-pulse operation results in a very effective correction of the charging background current. The charging-current contribution to the differential current is negligible, as described by

$$\Delta i_c \simeq -0.00567 C_i \, \Delta E m^{2/3} t^{-1/3} \quad (3.19)$$

where C_i is the integral capacitance. Such background contribution is smaller by more than an order of magnitude than the charging current of normal-pulse voltammetry. Accordingly, differential-pulse voltammetry allows measurements at concentrations as low as 10^{-8} M (about 1 µg/L). The improved detectability over DC polarography is demonstrated in Figure 3.7, which compares the response of both techniques for the antibiotic chloramphenicol present at the 1.3×10^{-5} M level. Similarly, the improvements over normal-pulse polarography are illustrated in Figure 3.8.

The peak-shaped response of differential-pulse measurements results also in improved resolution between two species with similar redox potentials. In various situations, peaks separated by 50 mV may be measured. Such quantitation depends not only on the corresponding peak potentials but also on the widths of the peak. The width of the peak (at half-height) is related to the electron stoichiometry:

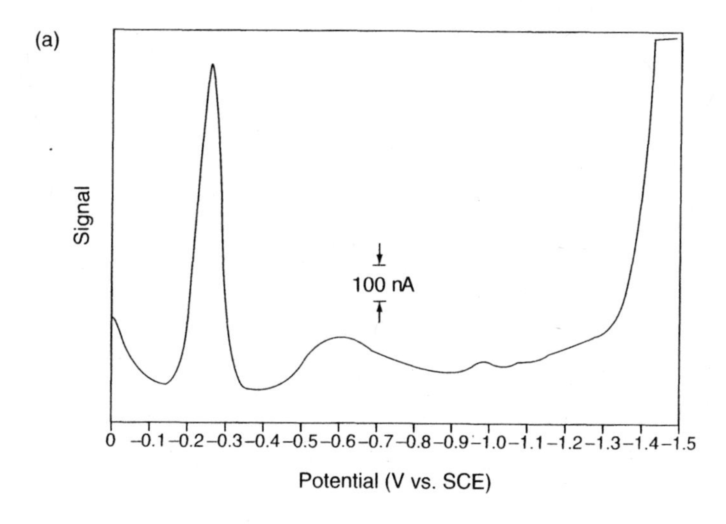

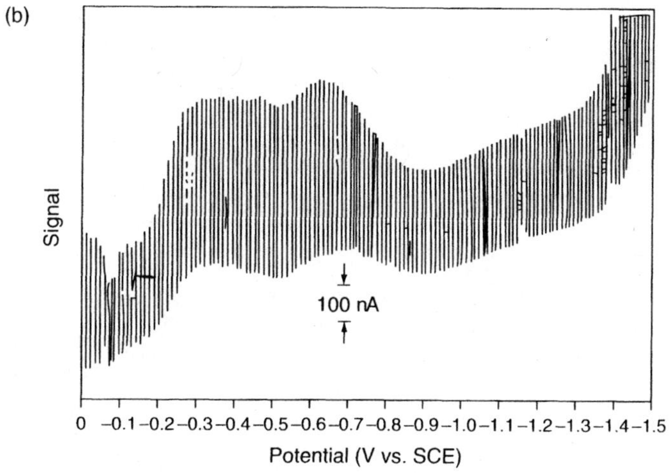

Figure 3.7 Differential-pulse (a) and DC (b) polarograms for a 1.3×10^{-5} M chloramphenicol solution. (Reproduced with permission from Ref. 7.)

$$W_{1/2} = \frac{3.52RT}{nF} \tag{3.20}$$

and thus corresponds to 30.1 mV for $n = 1$ (at 25°C). The peak-shaped response, coupled with the flat background current, makes the technique particularly useful for analysis of mixtures.

The selection of the pulse amplitude and potential scan rate usually requires a tradeoff among sensitivity, resolution, and speed. For example, larger pulse amplitudes result in larger and broader peaks. Pulse amplitudes of 25–50 mV,

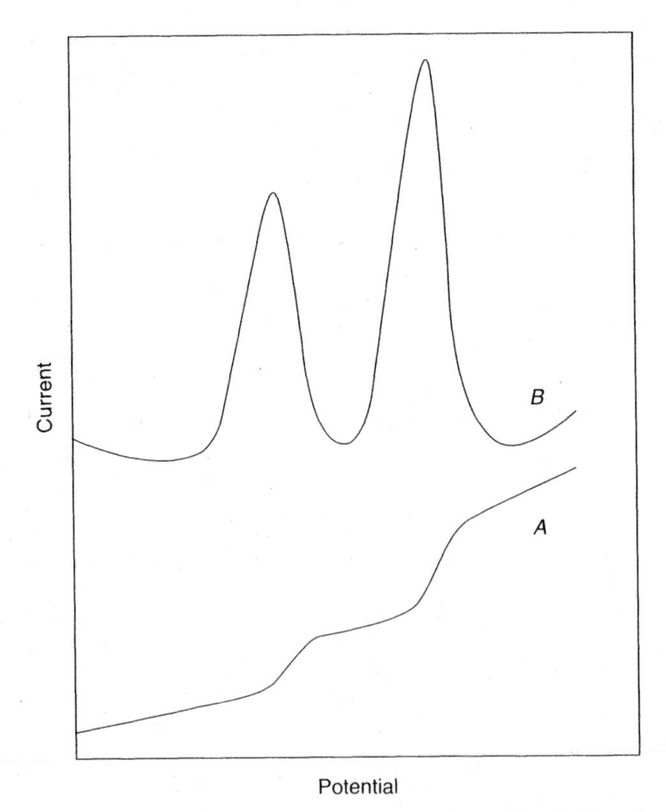

Figure 3.8 Normal-pulse (curve *A*) and differential-pulse (curve *B*) polarograms for a mixture of 1 mg/L cadmium and lead ions. Electrolyte, 0.1 M HNO_3.

coupled with a scan rate of 5 mV/s, are commonly employed. Irreversible redox systems result in lower and broader current peaks (i.e., inferior sensitivity and resolution) compared with those predicted for reversible systems (6). In addition to improvements in sensitivity and resolution, the technique can provide information about the chemical form in which the analyte appears (oxidation states, complexation, etc.).

3.3.3 Square-Wave Voltammetry

Square-wave voltammetry is a large-amplitude differential technique in which a waveform composed of a symmetric square wave, superimposed on a base staircase potential, is applied to the working electrode (8) (Fig. 3.9). The current is sampled twice during each square-wave cycle, once at the end of the forward pulse (at t_1) and once at the end of the reverse pulse (at t_2). Since the square-wave modulation amplitude is very large, the reverse pulses cause the reverse reaction of the product (of the forward pulse). The difference between the two measurements is plotted versus the base staircase potential.

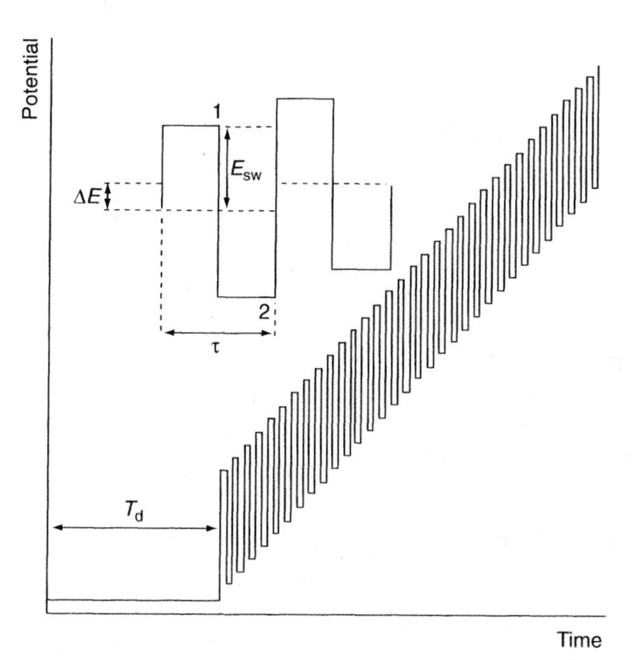

Figure 3.9 Square-wave waveform showing the amplitude E_{sw}, step height ΔE, square-wave period T, delay time T_d, and current measurement times 1 and 2. (Reproduced with permission from Ref. 9.)

A dimensionless plot of the theoretical forward, reverse, and difference currents is given in Figure 3.10 for a rapid reversible redox system. The resulting peak-shaped voltammogram is symmetric about the half-wave potential, and the peak current is proportional to the concentration. Excellent sensitivity accrues from the fact that the net current is larger than either the forward or reverse components (since it is the difference between them); the sensitivity is higher than that of differential pulse polarography (in which the reverse current is not used). Coupled with the effective discrimination against the charging background current, very low detection limits near 1×10^{-8} M can be attained. Comparison between square-wave and differential-pulse voltammetry for reversible and irreversible cases indicated that the square-wave currents are 4 and 3.3 times higher, respectively, than the analogous differential-pulse response (10). Figure 3.11 displays typical square-wave voltammograms obtained at a printed carbon strip electrode for increasing concentrations (1–10 ppm) of the nitroaromatic explosive 2,4,6-trinitrotoluene (TNT) (11).

The major advantage of square-wave voltammetry is its speed. The effective scan rate is given by $f\,\Delta E_s$. The term f is the square-wave frequency (in Hz) and ΔE_s is the step height. Frequencies of 1–100 cycles per second permit the use of extremely fast potential scan rates. For example, if $\Delta E_s = 10$ mV and $f = 50$ Hz, then the effective scan rate is 0.5 V/s. As a result, the analysis time is drastically reduced; a complete voltammogram can be recorded within a few

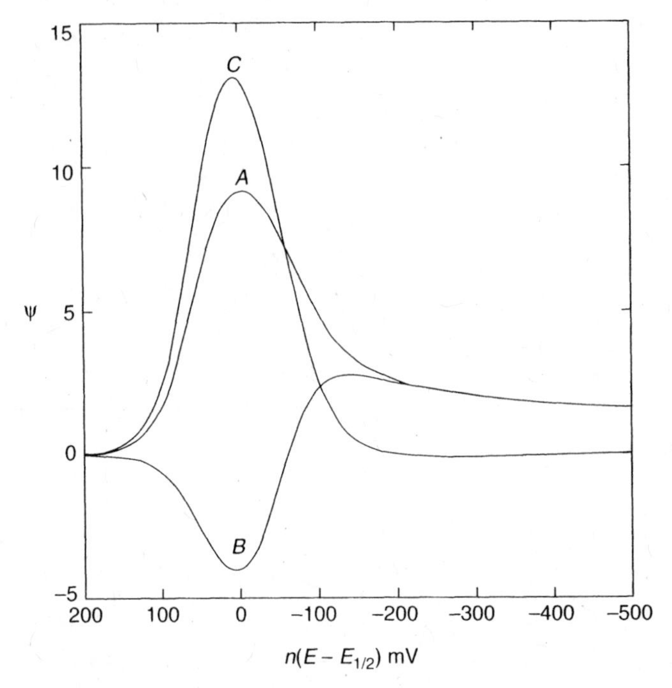

Figure 3.10 Square-wave voltammograms for reversible electron transfer: (curve *A*) forward current; (curve *B*) reverse current; (curve *C*) net current. (Reproduced with permission from Ref. 9.)

seconds, as compared with about 2–3 min in differential-pulse voltammetry. Because of the fast scan rates, the entire voltammogram is recorded on a single mercury drop. Hence, such an operation consumes many drops (compared to other pulse techniques). The inherent speed of square-wave voltammetry can greatly increase sample throughputs in batch (12) and flow (13) analytical operations. In addition, square-wave voltammetric detection for liquid chromatography and capillary electrophoresis can be used to resolve coeluting or comigrating species and assist in peak identification (14, 15). Kinetic studies can also benefit from the rapid scanning capability and the reversal nature of square-wave voltammetry.

3.3.4 Staircase Voltammetry

Staircase voltammetry has been proposed as a useful tool for rejecting the background charging current. The potential–time waveform involves successive potential steps of ~10 mV height and about 50 ms duration (Fig. 3.12). The current is sampled at the end of each step, where the charging current has decayed to a negligible value. Hence, this waveform couples the discrimination against the charging current with the experimental speed of linear scan voltammetry. Such an operation results in a peak-shaped current response, similar to that of linear scan experiments. Indeed, as the steps become smaller,

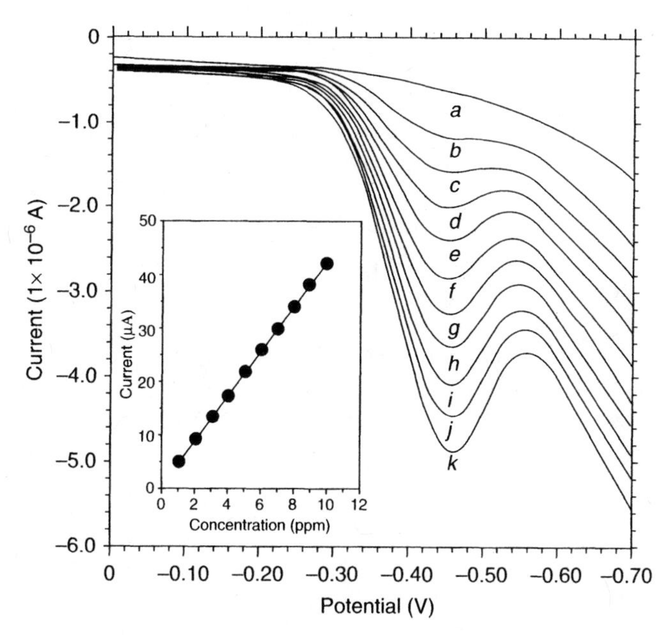

Figure 3.11 Square-wave voltammograms for TNT solutions of increasing concentration from 1 to 10 ppm (curves *b*–*k*), along with the background voltammogram (curve *a*) and resulting calibration plot (inset). (Reproduced with permission from Ref. 11.)

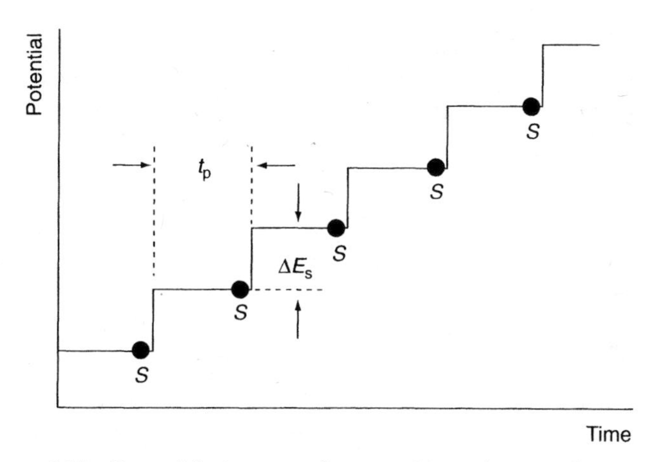

Figure 3.12 Potential–time waveform used in staircase voltammetry.

the equations for the staircase voltammetric response converge with those of linear scan voltammetry (16). As such, staircase voltammetry can be considered as the digital version of linear scan voltammetry. Similarly, cyclic staircase voltammetric experiments, in which the direction of the potential steps is reversed at a switching potential, result in a voltammetric response resembling cyclic voltammetry (but with a much reduced charging-current contribution).

3.4 AC VOLTAMMETRY

Alternating current (AC) voltammetry is a frequency-domain technique which involves the superimposition of a small amplitude AC voltage on a linear ramp (Fig. 3.13). Usually the alternating potential has a frequency of 50–100 Hz and an amplitude of 10–20 mV. The AC signal thus causes a perturbation in the surface concentration, around the concentration maintained by the DC potential ramp. The resulting AC current is plotted against the potential. Such a voltammogram shows a peak, the potential of which is the same as that of the polarographic half-wave potential. (At this region the sinusoid has maximum impact on the surface concentration, i.e., on the current.) For a reversible system, such a response is actually the derivative of the DC polarographic response. The height of the AC voltammetric peak is proportional to the concentration of the analyte and, for a reversible reaction, to the square root of the frequency (ω):

$$i_p = \frac{n^2 F^2 A \omega^{1/2} D^{1/2} C \Delta E}{4RT} \tag{3.21}$$

The term ΔE is the amplitude. The peak width is independent of the AC frequency, and is 90.4/n mV (at 25°C).

The detection of the ac component allows one to separate the contributions of the faradaic and charging currents. The former is phase-shifted by 45° relative to the applied sinusoidal potential, while the background component is 90° out of phase. The charging current is thus rejected using a phase-sensitive lock-in amplifier (able to separate the in-phase and out-of-phase current components). As a result, reversible electrode reactions yield a detection limit of ~5×10^{-7} M.

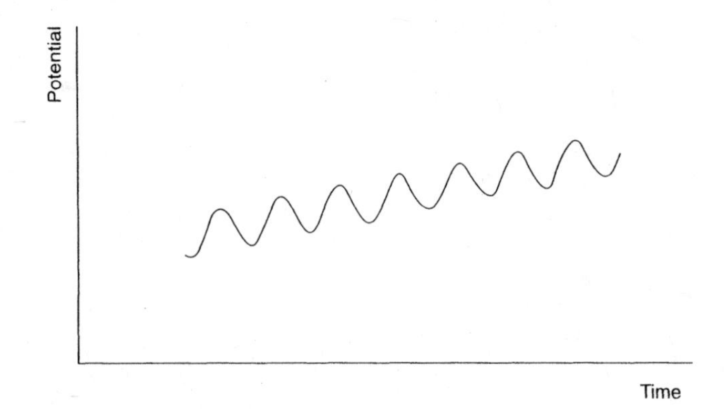

Figure 3.13 Potential-time waveform used in alternating current (ac) voltammetry.

Substantial loss in sensitivity is expected for analytes with slow electron transfer kinetics. This may be advantageous for measurements of species with fast electron transfer kinetics in the presence of one (e.g., dissolved oxygen) that is irreversible. (For the same reason, the technique is very useful for the study of electron processes.) Theoretical discussions on AC voltammetry are available in the literature (17–19).

Large-amplitude sine waves have been combined with conversion of the raw time domain (from the electrochemical cell) into the frequency domain (20,21). Such frequency-based sinusoidal voltammetry offers decoupling of the faradaic signal from the background components (i.e., lower detection limits), as well as generation of a distinct "fingerprint" frequency spectrum to aid in the identification of specific chemical molecules (20). Such decoupling of the faradaic signal reflects the fact that at large (i.e., >50 mV) amplitudes of the applied potential, it exhibits an intensity at higher order harmonics of the fundamental excitation frequency, as compared to the capacitive charging current that remains at the fundamental frequency. In addition to improved sensitivity at higher harmonics, redox species with different electrochemical properties can be detected selectively on the basis of their unique "fingerprint" frequency response (20,22). For example, measurements in the presence of dissolved oxygen have been accomplished on the basis of kinetic discrimination against the oxygen reduction process (22).

3.5 STRIPPING ANALYSIS

Stripping analysis is an extremely sensitive electrochemical technique for measuring trace metals (23,24). Its remarkable sensitivity is attributed to the combination of an effective preconcentration step with advanced measurement procedures that generates an extremely favorable signal : background ratio. Since the metals are preconcentrated into the electrode by factors of 100–1000, detection limits are lowered by two to three orders of magnitude compared to solution-phase voltammetric measurements. Hence, four to six metals can be measured simultaneously in various matrices at concentration levels down to 10^{-10} M, utilizing relatively inexpensive instrumentation. The ability to obtain such low detection limits strongly depends on the degree to which contamination can be minimized. Expertise in ultratrace chemistry is required.

Essentially, stripping analysis is a two-step technique. The first, or deposition step, involves the electrolytic deposition of a small portion of the metal ions in solution into the mercury electrode to preconcentrate the metals. This is followed by the stripping step (the measurement step), which involves the dissolution (stripping) of the deposit. Different versions of stripping analysis can be employed, depending on the nature of the deposition and measurement steps.

3.5.1 Anodic Stripping Voltammetry

Anodic stripping voltammetry (ASV) is the most widely used form of stripping analysis. In this case, the metals are being preconcentrated by electrodeposition into a small-volume mercury electrode (a thin mercury film or a hanging mercury drop). The preconcentration is done by cathodic deposition at a controlled time and potential. The deposition potential is usually 0.3–0.5 V more negative than E° for the least easily reduced metal ion to be determined. The metal ions reach the mercury electrode by diffusion and convection, where they are reduced and concentrated as amalgams:

$$M^{n+} + ne^- + Hg \rightarrow M(Hg) \tag{3.22}$$

The convective transport is achieved by electrode rotation or solution stirring (in conjunction with the mercury film electrode) or by solution stirring (when using the hanging mercury drop electrode). Quiescent solutions can be used when using mercury ultramicroelectrodes. The duration of the deposition step is selected according to the concentration level of the metal ions in question, from less than 0.5 min at the 10^{-7} M level to about 20 min at the 10^{-10} M level. The concentration of the metal in the amalgam, C_{Hg}, is given by Faraday's law

$$C_{Hg} = \frac{i_l t_d}{nFV_{Hg}} \tag{3.23}$$

where i_l is the limiting current for the deposition of the metal, t_d is the length of the deposition period, and V_{Hg} is the volume of the mercury electrode. The deposition current is related to the flux of the metal ion at the surface. The total amount of metal plated represents a small (and yet reproducible) fraction of the metal present in the bulk solution.

Following the preselected time of the deposition, the forced convection is stopped, and the potential is scanned anodically, linearly, or in a more sensitive potential–time (pulse) waveform that discriminates against the charging background current (usually square-wave or differential-pulse ramps). Such pulse excitations also offer reduced oxygen interferences and analyte replating, respectively. During this anodic scan, the amalgamated metals are reoxidized, stripped out of the electrode (in an order that is a function of each metal standard potential), and an oxidation (stripping) current is flowing:

$$M(Hg) \rightarrow M^{n+} + ne^- + Hg \tag{3.24}$$

Repetitive ASV runs can be performed with good reproducibility in connection to a short (30–60-s) "electrochemical cleaning" period at the final potential (e.g., +0.1 V using mercury electrodes). The potential–time sequence used in ASV, along with the resulting stripping voltammogram, is shown in Figure

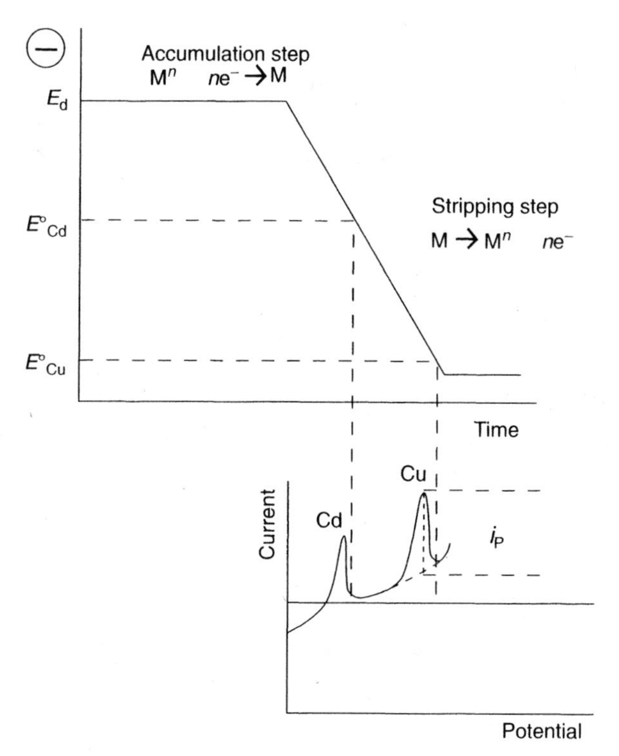

Figure 3.14 Anodic stripping voltammetry: the potential–time waveform (top), along with the resulting voltammogram (bottom).

3.14. The voltammetric peak reflects the time-dependent concentration gradient of the metal in the mercury electrode during the potential scan. Peak potentials serve to identify the metals in the sample. The peak current depends on various parameters of the deposition and stripping steps, as well as on the characteristics of the metal ion and the electrode geometry. For example, for a mercury film electrode, the peak current is given by

$$i_p = \frac{n^2 F^2 v^{1/2} A l C_{Hg}}{2.7RT} \tag{3.25}$$

where A and l are the area and thickness, respectively, of the film and v is the potential scan rate (during the stripping). The corresponding concentration profile in the film and nearby solution is displayed in Figure 3.15. (For very thin mercury films, diffusion in the film can be ignored and the peak current will be directly proportional to the scan rate.) For the hanging mercury drop, the following expression describes the stripping peak current:

$$i_p = 2.72 \times 10^5 n^{3/2} A D^{1/2} v^{1/2} C_{Hg} \tag{3.26}$$

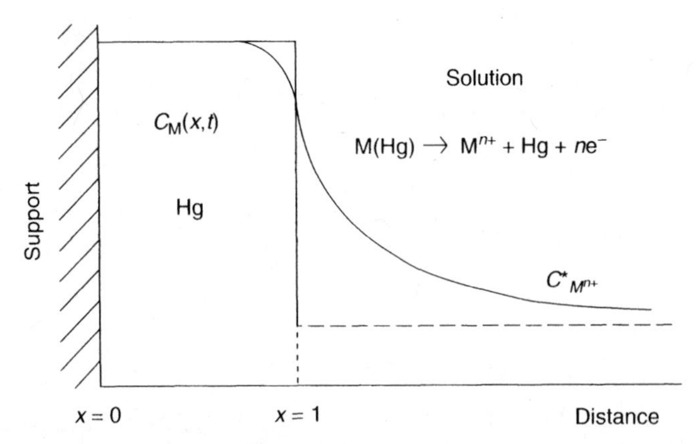

Figure 3.15 Concentration gradient of the metal in the mercury film electrode and nearby solution during the stripping step.

The mercury film electrode has a higher surface : volume ratio than does the hanging mercury drop electrode and consequently offers a more efficient preconcentration and higher sensitivity [Eqs. (3.23)–(3.26)]. In addition, the total exhaustion of thin mercury films results in sharper peaks and hence improved peak resolution in multicomponent analysis (Fig. 3.16). Because of the toxicity of mercury, considerable efforts have been devoted to the investigation of alternate electrode materials. Among the proposed alternative ("mercury-free") electrodes, the bismuth one offers the closest behavior to mercury (27,28). Both in-situ or preplated bismuth-film and bulk bismuth electrodes display a very attractive stripping behavior. Such behavior reflects the ability of bismuth to form "fused" multicomponent alloys with heavy metals. Gold electrodes have been particularly useful for stripping measurements of important trace metals (e.g., Hg, As, Se) with oxidation potentials more positive than those of mercury or bismuth. The combination of ultrasound with stripping voltammetry, has been proposed to allow the possibility of a wider use of solid electrodes and to overcome electrode passivation in real samples (29).

The major types of interferences in ASV procedures are overlapping stripping peaks caused by a similarity in the oxidation potentials (e.g., of the Pb, Tl, Cd, Sn or Bi, Cu, Sb groups), the presence of surface-active organic compounds that adsorb on the mercury electrode and inhibit the metal deposition, and the formation of intermetallic compounds (e.g., Cu–Zn), which affects the peak size and position. Knowledge of these interferences can lead to their prevention, through adequate attention to key operations.

Improved signal-to-background characteristics can be achieved using dual-working electrode techniques, such as ASV with collection or subtractive ASV (but at the expense of more complex instrumentation).

Other versions of stripping analysis, including potentiometric stripping, adsorptive stripping, and cathodic stripping schemes, have been developed to further expand its scope and power.

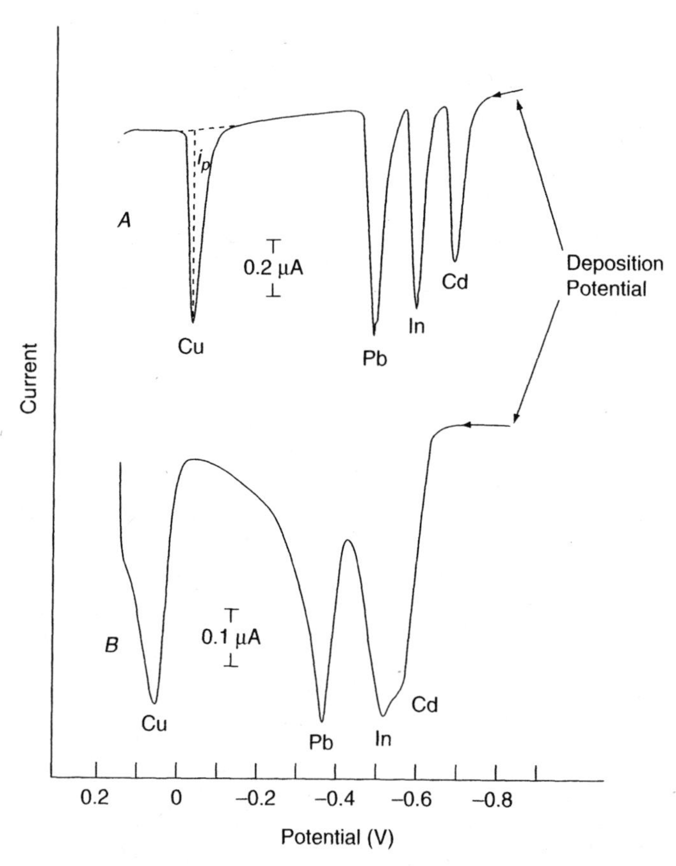

Figure 3.16 Stripping voltammograms for 2×10^{-7} M Cu^{2+}, Pb^{2+}, In^{3+}, and Cd^{2+} at the mercury film (*A*) and hanging mercury drop (*B*) electrodes. (Reproduced with permission from Ref. 25.)

3.5.2 Potentiometric Stripping Analysis

Potentiometric stripping analysis (PSA), known also as *stripping potentiometry*, differs from ASV in the method used for stripping the amalgamated metals (30). In this case, the potentiostatic control is disconnected following the preconcentration, and the concentrated metals are reoxidized by an oxidizing agent [e.g., O_2, Hg(II)] that is present in the solution:

$$M(\text{Hg}) + \text{oxidant} \rightarrow \text{M}^{n+} \tag{3.27}$$

A stirred solution is used also during the stripping step to facilitate the transport of the oxidant. Alternately, the oxidation can be carried out by passing a constant anodic current through the electrode. During the oxidation step, the variation of the working electrode potential is recorded, and a stripping curve,

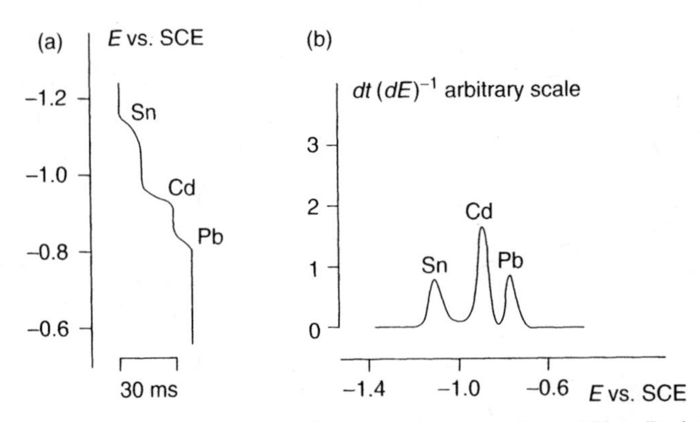

Figure 3.17 Stripping potentiograms for a solution containing 100 µg/L tin, cadmium, and lead, with 80-s accumulation at −1.40 V. (Reproduced with permission from Ref. 30.)

like the one shown in Figure 3.17a, is obtained. When the oxidation potential of a given metal is reached, the potential scan is slowed down as the oxidant (or current) is used for its stripping. A sharp potential step thus accompanies the depletion of each metal from the electrode. The resulting potentiogram thus consists of stripping plateaus, as in a redox titration curve. The transition time needed for the oxidation of a given metal t_M is a quantitative measure of the sample concentration of the metal

$$t_M \propto C_{M^{n+}} t_d / C_{ox} \tag{3.28}$$

where C_{ox} is the concentration of the oxidant. Hence, the signal may be increased by decreasing the oxidant concentration. The qualitative identification relies on potential measurements (in accordance with the Nernst equation for the amalgamated metal):

$$E = E^\circ + \frac{RT}{nF} \ln[M^{n+}]/M(Hg) \tag{3.29}$$

where E° denotes the standard potential for the redox couple $M^{n+}/M(Hg)$.

Modern PSA instruments use microcomputers to register fast stripping events and to convert the wave-shaped response to a more convenient peak over a flat baseline. Such differential display of *dt*/*dE* versus *E* is shown in Figure 3.17b. The use of nondeaerated samples represents an important advantage of PSA (over analogous ASV schemes), particularly in field applications. In addition, such potential–time measurements eliminate the need for amplification when microelectrodes are concerned. By obviating the need for stirring or deoxygenating the solution, the coupling of PSA with microelectrodes permits convenient trace analysis of very small (5-µL) samples. PSA is also less susceptible to interfering surfactant effects, and hence can simplify the pretreatment of biological samples. A more detailed treatment of the theoretical foundation of PSA is given in Ref. 30.

About 20 amalgam-forming metals, including Pb, Sn, Cu, Zn, Cd, Bi, Sb, Tl, Ga, In, and Mn, are easily measurable by stripping strategies (ASV and PSA) based on cathodic deposition onto mercury electrodes. Additional metals, such as Se, Hg, Ag, Te, and As, are measurable at bare solid electrodes such as carbon or gold.

3.5.3 Adsorptive Stripping Voltammetry and Potentiometry

Adsorptive stripping analysis greatly enhances the scope of stripping measurements toward numerous trace elements (31–33). This relatively new strategy involves the formation, adsorptive accumulation, and reduction of a surface-active complex of the metal (Fig. 3.18). Both voltammetric and potentiometric stripping schemes, with a negative-going potential scan or constant cathodic current, respectively, can be employed for measuring the adsorbed complex. Most procedures involve the reduction of the metal in the adsorbed complex (although it is also possible to exploit the reduction of the ligand). The response of the surface-confined species is directly related to its surface concentration, with the adsorption isotherm (commonly that of Langmuir—discussed in Section 2.1), providing the relationship between the surface and bulk concentrations of the adsorbate. As a result, calibration curves display nonlinearity at high concentrations. The maximum adsorption density is related to the size and surface concentration of the adsorbed complex.

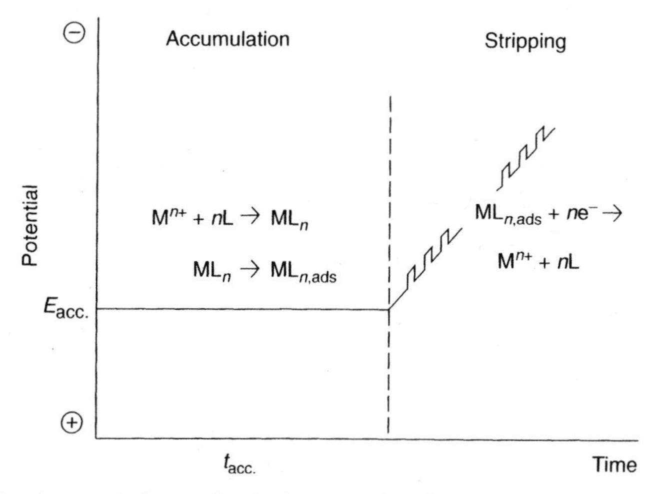

Figure 3.18 Accumulation and stripping steps in adsorptive stripping measurements of a metal ion (M^{n+}) in the presence of an appropriate chelate agent (L).

Short adsorption times (1–5 min) result in a very effective interfacial accumulation. The reduction step is also very efficient as the entire collected complex is reduced. Such a combination thus results in extremely low detection limits (10^{-10}–10^{-11} M) for important metals, including uranium, vanadium, aluminum, or molybdenum. Even lower levels, for example, 10^{-12} M of platinum, titanium, cobalt, chromium, or iron, can be measured by coupling the adsorption accumulation with catalytic reactions. In this case, the response of the accumulated complex is greatly amplified through a catalytic cycle, for example, in the presence of an oxidant. The adsorptive approach may also offer improvements in selectivity or sensitivity for metals (e.g., tin, nickel) that are measurable also by conventional stripping analysis. Examples of adsorptive stripping schemes for measuring trace metals are listed in Table 3.2. All procedures rely on a judicious choice of chelating agent. The resulting complex should be surface active and electroactive; in addition, selective complexation can be used to enhance the overall selectivity.

Besides trace metals, adsorptive stripping voltammetry has been shown to be highly suitable for measuring organic compounds (including cardiac or anticancer drugs, nucleic acids, vitamins, or pesticides) that exhibit surface-active properties. Depending on their redox activity, the quantitation of the adsorbed organic compounds may proceed, through oxidation or reduction. For example, modern adsorptive stripping voltammetric and potentiometric methods represent highly sensitive tools for detecting ultratrace levels of nucleic acids. Figure 3.19 displays the adsorptive stripping potentiometric

TABLE 3.2 Common Adsorptive Stripping Schemes for Measurements of Trace Metals

Metal	Complexing Agent	Supporting Electrolyte	Detection Limit (M)	Ref.
Al	Di(hydroxyanthraquinone sulfonic acid)	BES buffer	1×10^{-9}	34
Be	Thorin	Ammonia buffer	3×10^{-9}	35
Co	Nioxime	HEPES buffer	6×10^{-12}	36
Cr	Diethylenetriamine-pentaacetic acid	Acetate buffer	4×10^{-10}	37
Fe	Solochrome violet RS	Acetate buffer	7×10^{-10}	38
Mn	Eriochrome Black T	PIPES buffer	6×10^{-10}	39
Mo	Oxine	Hydrochloric acid	1×10^{-10}	40
Ni	Dimethylglyoxime	Ammonia buffer	1×10^{-10}	41
Pt	Formazone	Sulfuric acid	1×10^{-12}	42
Sn	Tropolone	Acetate buffer	2×10^{-10}	43
Ti	Mandelic acid	Potassium chlorate	7×10^{-12}	44
U	Oxine	PIPES buffer	2×10^{-10}	45
V	Catechol	PIPES buffer	1×10^{-10}	46

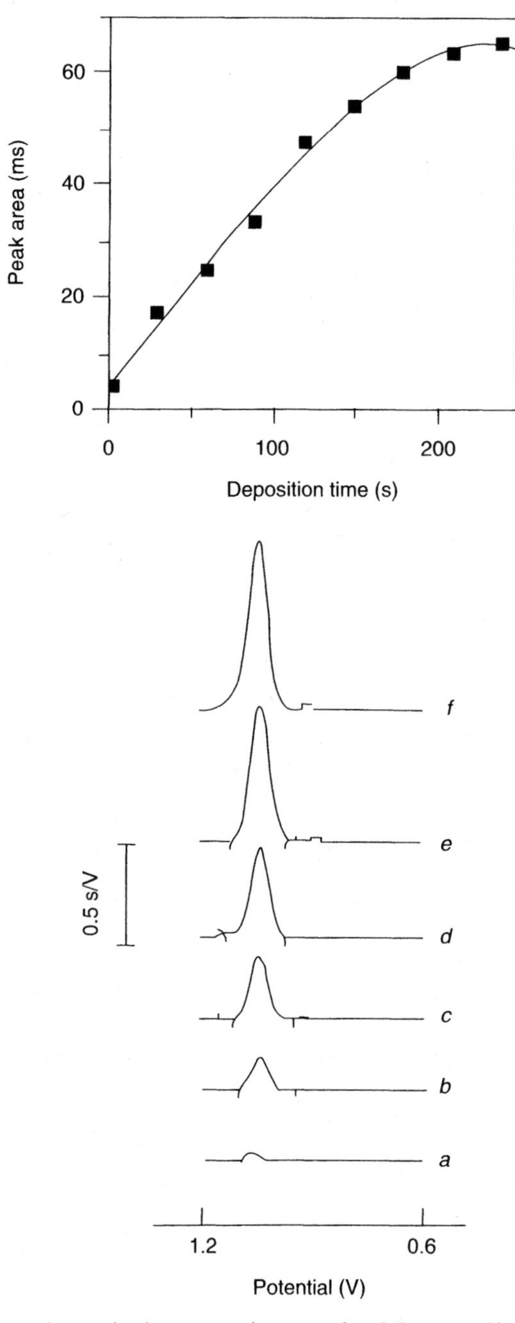

Figure 3.19 Adsorptive stripping potentiograms for 0.5 ppm calf thymus DNA following different adsorption times: 1–150 s (*a–f*). (Reproduced with permission from Ref. 47.)

response of the carbon paste electrode for 0.5 ppm calf thymus DNA following different adsorption times. Nonelectroactive macromolecules may also be determined following their interfacial accumulation from tensammetric peaks (resulting from their adsorption–desorption mechanism).

3.5.4 Cathodic Stripping Voltammetry

Cathodic stripping voltammetry (CSV) is the mirror image of ASV. It involves anodic deposition of the analyte, followed by stripping in a negative-going potential scan:

$$A^{n-} + Hg \underset{\text{stripping}}{\overset{\text{deposition}}{\rightleftharpoons}} HgA + ne^- \tag{3.30}$$

The resulting reduction peak current provides the desired quantitative information. Cathodic stripping voltammetry is used to measure a wide range of organic and inorganic compounds, capable of forming insoluble salts with mercury. Among these are various thiols or penicillins, as well as halide ions, cyanide, and sulfide. Highly sensitive measurements can thus be performed, as illustrated in Figure 3.20 for the direct determination of subnanomolar concentrations of iodide in seawater.

Anions (e.g., halides) that form insoluble silver salts can be measured also at a rotating silver disk electrode. In this, the deposition and stripping steps involve the following reaction:

$$Ag + X^- \rightleftharpoons AgX + e^- \qquad X^- = Cl^-, Br^- \tag{3.31}$$

Copper-based electrodes can also be employed for the same task.

3.5.5 Abrasive Stripping Voltammetry

Abrasive striping voltammetry, introduced by Scholz' team (49), provides a qualitative and quantitative analysis of solid materials. The method involves a mechanical transfer (by rubbing) of trace amounts of a solid sample onto the electrode surface (usually a paraffin-impregnated graphite electrode), followed by voltammetric measurement and stripping of the accumulated material. After the voltammetric measurement, the surface is "cleaned" by rubbing it onto a smooth filter paper. The technique has been shown useful for different aspects of solid-state analysis, including fingerprint identification of alloys, study of minerals, analysis of pigments or pesticides, and fundamental investigations of electrode processes of solid compounds.

3.5.6 Applications

The remarkable sensitivity, broad scope, and low cost of stripping analysis have led to its application in a large number of analytical problems. As illustrated

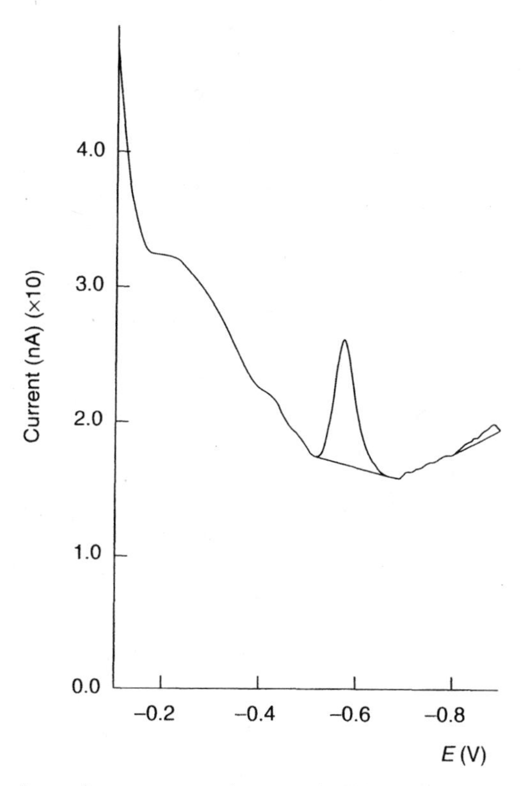

Figure 3.20 Stripping voltammograms for trace iodide in seawater. (Reproduced with permission from Ref. 48.)

in Figure 3.21, over 30 trace elements can be conveniently measured in various matrices by the various versions of stripping analysis. The technique has thus proved useful for the determination of numerous trace metals in environmental, industrial, and clinical samples, as well as for assays of foodstuffs, beverages, gunshot residues, and pharmaceutical formulations. Selected applications are listed in Table 3.3. Figure 3.22 displays adsorptive stripping voltammograms for chromium in various environmental (soil and groundwater) samples from contaminated nuclear energy sites. Stripping analysis has also been an important technique for screening for blood lead in children (58) and for monitoring arsenic in various water samples (67). Many other unique applications of stripping analysis, including studies of metal speciation (oxidation state, metal–ligand interactions) in natural waters, on-line monitoring of industrial processes, in situ oceanographic surveys, or remote environmental sensing, have been reported and reviewed (50,68–70). The technique has also been extremely useful for monitoring metal tags (including nanoparticle tracers) in connection with bioaffinity assays of DNA and proteins (see Ref. 71 and Chapter 6, below).

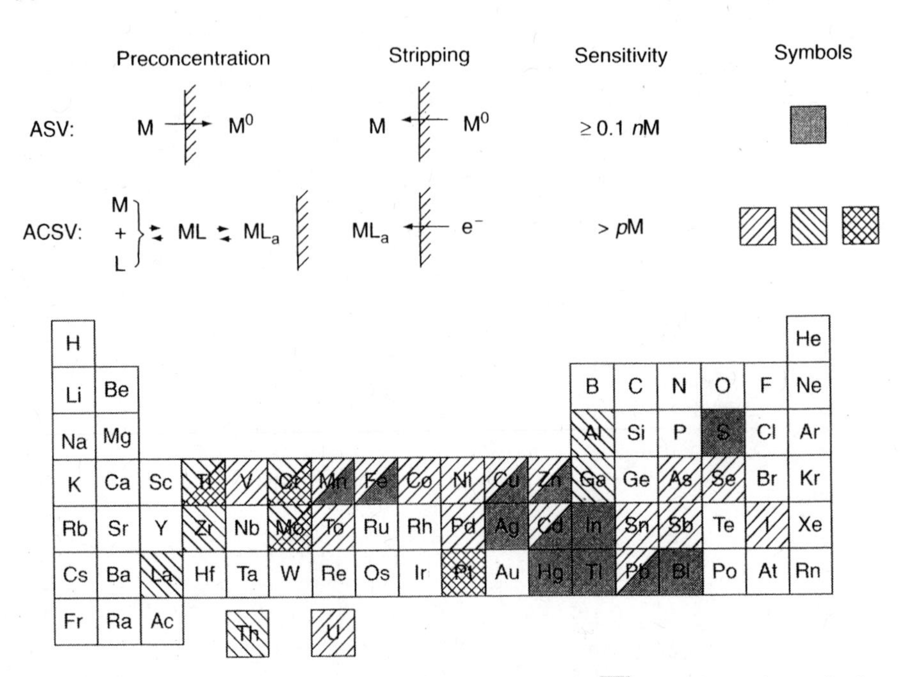

Figure 3.21 Elements measured by conventional ASV [dark shaded box] and adsorptive stripping schemes with reduction of the element in the complex [hatched box], reduction of the ligand [hatched box], or a catalytic process [cross-hatched box]. (Reproduced with permision from Ref. 50.)

TABLE 3.3 Representative Applications of Stripping Analysis

Metal	Sample Matrix	Stripping Mode	Working Electrode	Ref.
Antimony	Gunshot residue	ASV	MFE	51
Cadmium	Lake water	ASV	MFE	52
Chromium	Soil	AdSV	HMDE	53
Cobalt	Soil	AdSV	Bismuth	54
Copper	Steel	ASV	HMDE	55
Iodide	Seawater	CSV	HMDE	48
Iron	Wine	AdSV	HMDE	56
Iron	Seawater	Catalytic AdSV	HMDE	57
Lead	Blood	PSA	MFE	58
Lead	Paint	ASV	—	59
Mercury	Fish	ASV	Au	60
Nickel	Plant leaves	AdSV	HMDE	61
Platinum	Gasoline	AdSV	HMDE	62
Selenium	Soil	CSV	HMDE	63
Thallium	Urine	ASV	HMDE	64
Titanium	Seawater	AdSV	HMDE	44
Uranium	Groundwater	AdSV	HMDE	65
Zinc	Eye tissue	ASV	HMDE	66

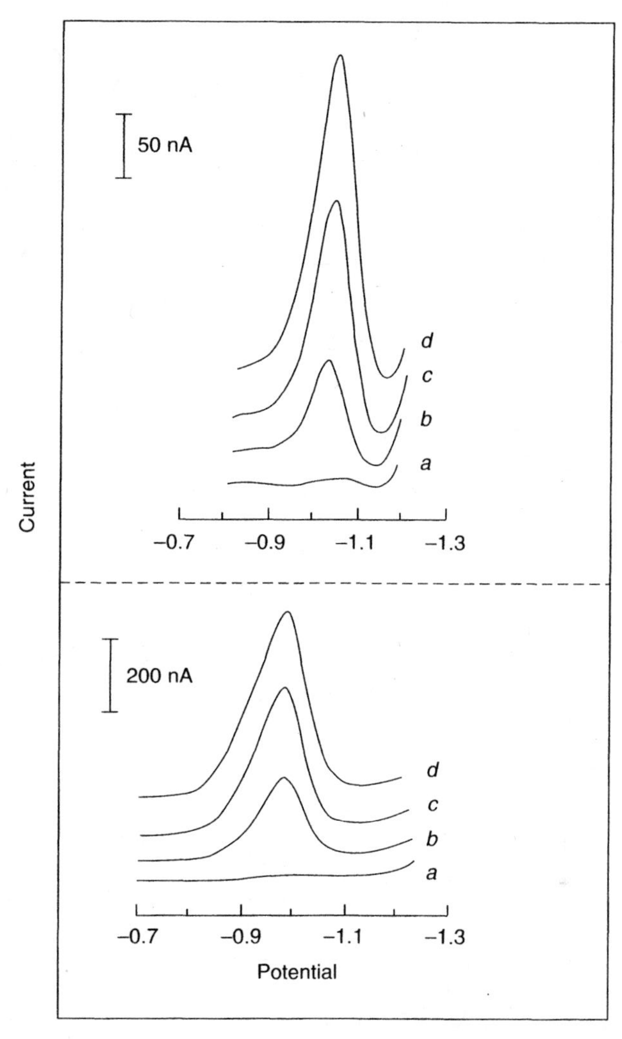

Figure 3.22 Adsorptive stripping voltammograms of chromium in groundwater (top) and soil (bottom) samples, using cupferron as a chelating agent. Top: curve *a*, response for electrolyte; curve *b*, same as curve *a* but after spiking 20 μL of the sample (500-fold dilution); curves *c* and *d*, same as curve *b*, but after additions of 0.1 μg/L chromium; 20 s adsorption. Bottom: curve *a*, response for the electrolyte; curve *b*, same as curve *a* but after spiking 5 μL of the soil extract (2000-fold dilution); curves *c* and *d*, same as curve *b* but after additions of 0.5 μ/L chromium; 15 s adsorption. (Reproduced with permission from Ref. 53.)

3.6 FLOW ANALYSIS

An electrochemical detector uses the electrochemical properties of target analytes for their determination in a flowing stream. While parameters such as current, potential, conductivity, and capacitance can be monitored by various electrochemical detectors, our discussion will focus primarily on the most popular constant-potential measurements. Controlled-potential detectors are ideally suited for monitoring analytes that are electroactive at modest potentials. Such devices are characterized by a remarkable sensitivity (down to the picogram level), high selectivity (toward electroactive species), wide linear range, low dead volumes, fast response, and relatively simple and inexpensive instrumentation. Electrochemical detectors are commonly used in many clinical, environmental, and industrial laboratories in connection with automated flow systems (e.g., flow injection analyzers) or separation techniques [particularly liquid chromatography, conventional and microchip capillary zone electrophoresis (CZE) and on-line microdialysis]. Such coupling of electrochemical detectors with advanced separation steps allows electroanalysis to address highly complex samples.

3.6.1 Principles

Electrochemical detection is usually performed by controlling the potential of the working electrode at a fixed value (corresponding to the limiting current plateau region of the compounds of interest) and monitoring the current as a function of time. The current response thus generated reflects the concentration profiles of these compounds as they pass through the detector. Hence, detection for liquid chromatography or flow injection systems results in sharp current peaks (reflecting the passage of the eluted analyte or sample zone, respectively). Accordingly, the magnitude of the peak current serves as a measure of the concentration. Typical response peaks recorded during an automated flow injection operation are displayed in Figure 3.23. The current peaks are superimposed on a constant background current (caused by redox reactions of the mobile phase or carrier solutions). Larger background currents, expected at high potentials, result in increased (flow rate-dependent) noise level. In particular, the cathodic detection of reducible species is hampered by the presence of traces of oxygen in the flowing solution. Such background noise is strongly influenced by the pulsation of the pump.

The applied potential affects not only the sensitivity and signal : noise characteristics, but also the selectivity of amperometric measurements. In general, a lower potential is more selective and a higher one, more universal. Thus, compounds undergoing redox potentials at lower potentials can be detected with greater selectivity. For example, Figure 3.24 displays electropherograms for a beer sample recorded at different detection potentials. Far fewer peaks are seen in the lower-potential (850 mV) electropherogram. However, the detection potential must be sufficient high to oxidize (or reduce) the com-

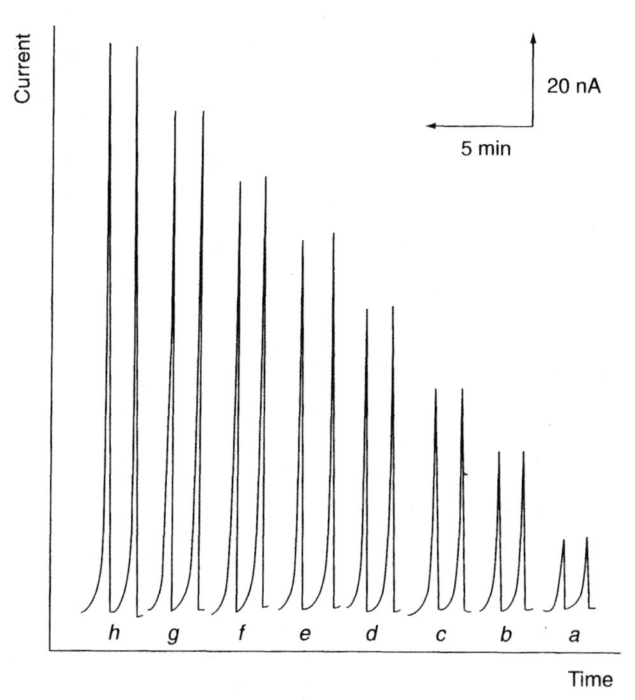

Figure 3.23 Typical amperometric (readout during automated flow injection assays of ethanol solutions of increasing concentrations in 2×10^{-5} M steps at a carbon paste enzyme electrode detector. Curves *a–h*: 2×10^{-5} M – 1.6×10^{-4} M ethanol.

pounds of interest. Selection of the applied potential relies on the construction of hydrodynamic voltammograms. These can be obtained by making repeated flow injections of the analyte solution while recording the current at different potentials. The resulting voltammogram has a characteristic wave (sigmoidal) shape. Although it is common to operate the detector on the limiting-current plateau region, a lowering of the operating potential (to the rising portion of the wave) can be used to improve the selectivity and lower the detection limit. Comparison of hydrodynamic voltammograms for the sample and standard solutions can provide important qualitative information.

Depending on their conversion efficiency, electrochemical detectors can be divided into two categories: those that electrolyze only a negligible fraction (0.1–5%) of the electroactive species passing through the detector (amperometric detectors), and those for which the conversion efficiency is approaching 100% (coulometric detectors). Unfortunately, the increased conversion efficiency of the analyte is accompanied by a similar increase for the electrolyte (background) reactions, and no lowering of detection limits is realized.

Various (pre- or postcolumn) chemical and biochemical derivatization schemes have thus been employed for expanding the scope of electrochemical detectors toward electroinactive analytes (through the introduction of elec-

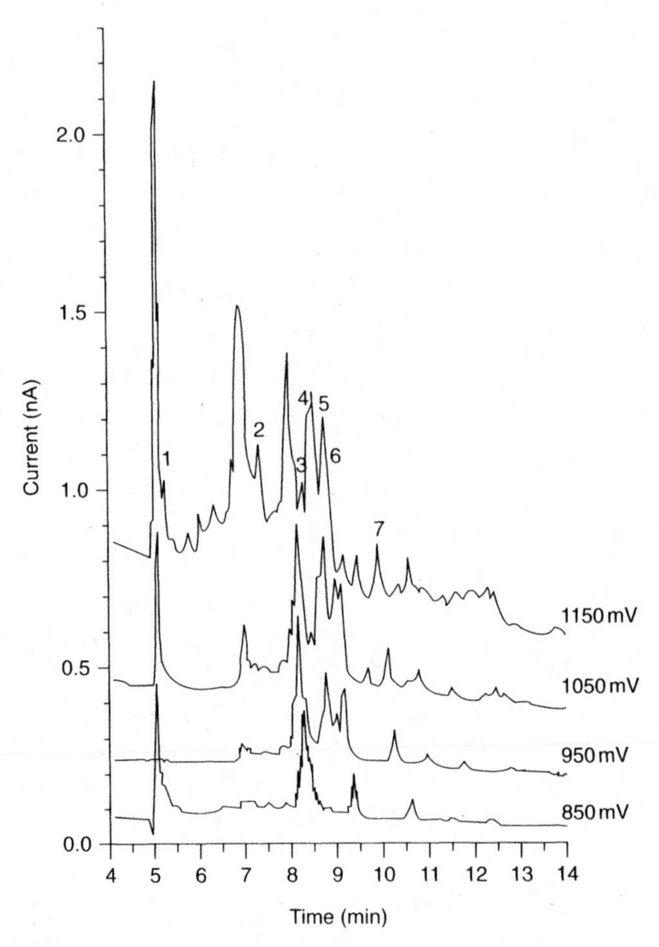

Figure 3.24 Capillary electrophoresis/amperometric response of a Bud Light beer sample using different detection potentials. (Reproduced with permission from Ref. 72.)

troactive functionalities) and for improving the detection of species with normally unfavorable redox properties. Post-column addition can also be used to adjust the conditions (e.g., pH) to meet the needs of the electrochemical detection.

3.6.2 Cell Design

A wide range of cell designs have been used for electrochemical monitoring of flowing streams. The cell design must fulfill the requirements of high signal : noise ratio, low dead volume, well-defined hydrodynamics, small ohmic drop, and ease of construction and maintenance (polishing). In addition, the reference and counter electrodes should be located on the downstream side of the working electrode, so that reaction products at the counter electrode or leakage from the reference electrode do not interfere with the working electrode. The distance between the column outlet and the working electrode affects the post-separation band broadening.

The most widely used amperometric detectors are based on the thin-layer and wall-jet configurations (Fig. 3.25). The thin-layer cell relies on a thin layer of solution that flows parallel to the planar electrode surface, which is embedded in a rectangular channel. The flow channel is formed by two plastic blocks pressing a thin Teflon gasket, which defines the very small dead volume (~1 μL). In the wall-jet design, the stream flows from a nozzle perpendicularly onto a flat electrode surface (the wall), and then spreads radially over the surface. The electrode diameter is significantly larger than the nozzle inlet. Since the jet remains intact up to quite large inlet electrode separations, it is possible to employ also large-volume wall-jet detectors that offer decreased sensitivity to the properties of the mobile phase, and a simplified fabrication. Both the thin-layer and wall-jet designs commonly rely on disk working electrodes made of carbon (e.g., glassy carbon, diamond, or paste) or metal (such as gold or platinum).

It is also possible to employ detectors with solutions flowing over a static mercury drop electrode or a carbon fiber microelectrode, or to use flow-through electrodes, with the electrode simply an open tube or porous matrix. The latter can offer complete electrolysis, namely, coulometric detection. The extremely small dimensions of ultramicroelectrodes (discussed in Section 4.5.4) offer the advantages of flow-rate independence (and hence a low noise level) and operation in nonconductive mobile phases (such as those of normal-phase chromatography or supercritical fluid chromatography).

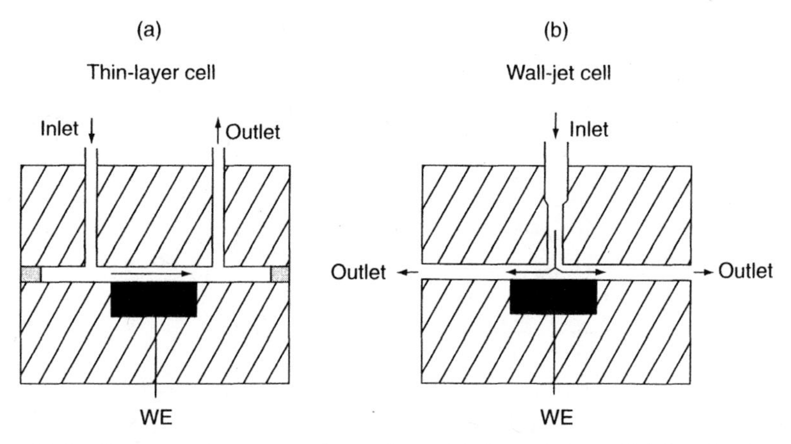

Figure 3.25 Common detector configurations: thin-layer (channel) (a) and wall-jet (b) flow cells.

Ultramicroelectrodes can also greatly benefit modern microseparation techniques such as open-tube liquid chromatography or capillary-zone electrophoresis (CZE) (73). For example, cylinder-shaped carbon or copper fibers can be inserted into the end of the CE separation capillary (e.g., see Fig. 3.26). Such alignment of the working electrode with the end of the capillary represents a challenge in combining electrochemistry with CZE.

Capillary-zone electrophoresis has more recently established itself as an important separation tool, due to its impressive separation power. Since CZE separations rely on the application of strong electric fields for separating the analytes, it is essential to isolate the low detection potential from the high voltage (10–30 kV) used to effect the separation (75). This can be accomplished by using a decoupling device (e.g., Nafion or cellulose acetate joints, porous glass) or via an end-column detection (i.e., placement of the detector opposite to the capillary outlet). The latter relies on the dramatic drop of the potential across small capillaries (of ≤25 μm). Figure 3.27 depicts a typical end-column electropherogram for femtomole quantities of dopamine, isoproterenol, and catechol. Since the sensitivity of electrochemical detection is not compromised by the low volumes used in CZE systems, extremely low mass detection limits (in the attomole range) can be obtained. Such high sensitivity toward easily oxidizable or reducible analytes rivals that of laser-induced fluorescence [which is currently (as of 2005) the method of choice for most CZE applications], and makes CZE/electrochemistry an ideal tool for assays of many small-volume samples.

Electrochemical detection offers also great promise for CZE microchips, and for other chip-based analytical microsystems (e.g., "Lab-on-a-Chip) discussed in Section 6.3 (77–83). Particularly attractive for such microfluidic devices are the high sensitivity of electrochemical detection, its inherent miniaturization of both the detector and control instrumentation, low cost, low power demands, and compatibility with micromachining technologies. Various detector configurations, based on different capillary/working-electrode

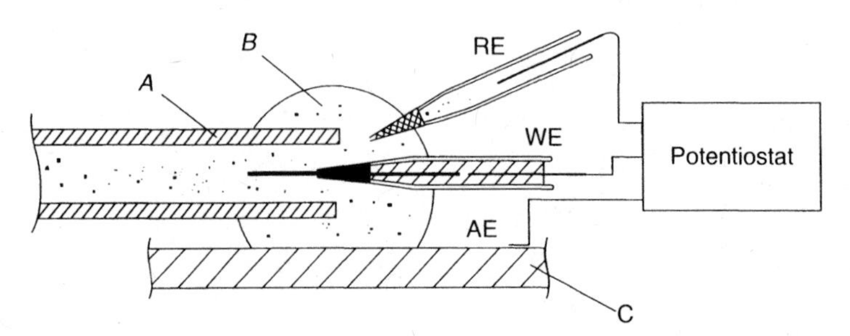

Figure 3.26 Schematic of a carbon fiber amperometric detector for capillary electrophoresis: *A*, fused-silica capillary; *B*, eluent drop; *C*, stainless-steel plate; RE, reference electrode; WE, working electrode, AE, auxiliary electrode. (Reproduced with permission from Ref. 74.)

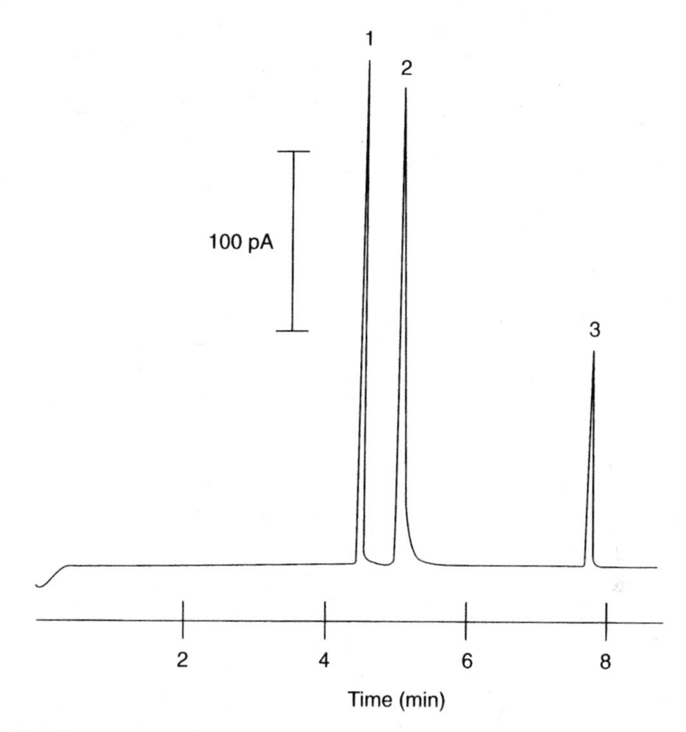

Figure 3.27 Electrophoretic separation of catechols with an end-column detection. Detection potential, +0.8 V; separation voltage, 20 kV. The peaks correspond to 4.6 fmol dopamine (1), 4.1 fmol isoproterenol (2), and 2.7 fmol catechol (3). (Reproduced with permission from Ref. 76.)

arrangements and the position of the electrode relative to the flow direction, have been proposed. These include flow-by (80), flow-onto (81), and flow-through (82) configurations (Fig. 3.28). In-channel detection (i.e., placement of a flow-by working electrode within the separation channel), which obviates postcapillary band dispersion effects, is also possible but usually requires an electrical decoupler that isolates the detector from the separation voltage (83).

3.6.3 Mass Transport and Current Response

Well-defined hydrodynamic conditions, with high rate of mass transport, are essential for a successful use of electrochemical detectors. According to the Nernst approximate approach, the thickness of the diffusion layer (δ) is empirically related to the solution flow rate (U) via

$$\delta = B/U^{\alpha} \tag{3.32}$$

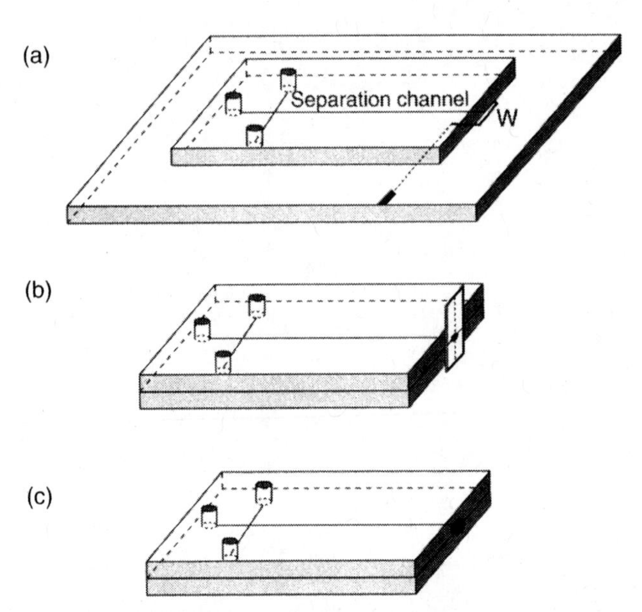

Figure 3.28 Common configurations of electrochemical detectors for CE microchips, based on different capillary/working-electrode arrangements and the position of the electrode (w) relative to the flow direction: (a) flow by (using two plates); (b) flow onto (with the surface normal to the flow direction); (c) flow through (with the detector placed directly on the channel exit). (Reproduced with permission from Ref. 78.)

where B and α are constants for a given set of conditions, with α ranging between 0.33 and 1.0. By substituting Eq. (3.32) in the general current response for mass-transport-controlled reactions [$i_l = (nFADC/\delta)$], one obtains the limiting steady-state response of flow-through electrodes

$$i_l = nFAK_m CU^{\alpha} \tag{3.33}$$

where K_m is the mass transport coefficient (D/B).

A more rigorous treatment takes into account the hydrodynamic characteristics of the flowing solution. Hence, expressions for the limiting currents (under steady-state conditions) have been derived for various electrode geometries by solving the three-dimensional convective diffusion equation:

$$\frac{\partial C}{\partial t} = D\left(\frac{\partial^2 C}{\partial x^2} + \frac{\partial^2 C}{\partial y^2} + \frac{\partial^2 C}{\partial z^2}\right) - \left(U_x \frac{\partial C}{\partial x} + U_y \frac{\partial C}{\partial y} + U_z \frac{\partial C}{\partial z}\right) \tag{3.34}$$

The resulting equations, arrived at by setting appropriate initial and boundary conditions (depending on the particular electrode), are given in Table 3.4.

TABLE 3.4 Limiting-Current Response of Various Flow-Through Electrodes[a]

Electrode Geometry	Limiting-Current Equation
Tubular	$i = 1.61nFC(DA/r)^{2/3}U^{1/3}$
Planar (parallel flow)	$i = 0.68nFCD^{2/3}\nu^{-1/6}(A/b)^{1/2}U^{1/2}$
Thin-layer cell	$i = 1.47nFC(DA/b)^{2/3}U^{1/3}$
Planar (perpendicular)	$i = 0.903nFCD^{2/3}\nu^{-1/6}A^{3/4}u^{1/2}$
Wall-jet detector	$i = 0.898nFCD^{2/3}\nu^{-5/12}a^{-1/2}A^{3/8}U^{3/4}$

[a] *Definition of terms*: a = diameter of inlet, A = electrode area, b = channel height, C = concentration (mM), F = Faraday constant, D = diffusion coefficient, ν = kinematic viscosity, r = radius of tubular electrode, U = average volume flow rate, u = velocity (cm/s), n = number of electrons.
Source: Adapted from Ref. 84.

A generalized equation for the limiting current response of different detectors, based on the dimensionless Reynolds (Re) and Schmidt (Sc) numbers, has been derived by Hanekamp and coworkers (84)

$$i_l = nkFCD(\mathrm{Sc})^{\beta}b(\mathrm{Re})^{\alpha} \tag{3.35}$$

where k is a dimensionless constant and b is the characteristic electrode width.

In the case of coulometric detectors (with complete electrolysis), the limiting current is given by Faraday's law:

$$i_l = nFCU \tag{3.36}$$

3.6.4 Detection Modes

The simplest, and by far the most common, detection scheme is measurement of the current at a constant potential. Such fixed-potential amperometric measurements have the advantage of being free of double-layer charging and surface transient effects. As a result, extremely low detection limits—on the order of 1–100 pg ($\sim 10^{-14}$ mol of analyte)—can be achieved. In various situations, however, it may be desirable to change (scan, pulse, etc.) the potential during the detection.

Potential-scanning detectors can increase the information content over that of fixed-potential operation. By rapidly recording numerous voltammograms during the elution, one obtains a three-dimensional detector response of the current against potential and time. Such addition of the redox potential selectivity can offer immediate identification of eluting peaks, and helps resolve chronomatographically coeluting components. Different approaches to swept-potential detectors based on square-wave voltammetry (14,15), frequency-based sinusoidal voltammetry (21), or phase-sensitive AC voltammetry (85) have been reported. The greater selectivity of potential-scanning detection is accompanied by higher detection limits (vs. fixed-potential amperometry), because of the additional background current associated with the potential scan.

Pulsed amperometric detection (PAD), introduced by Johnson and LaCourse (86,87) has greatly enhanced the scope of liquid chromatography/electrochemistry (88). This detection mode overcomes the problem of lost activity of noble metal electrodes associated with the fixed-potential detection of compounds such as carbohydrates, alcohols, amino acids, or aldehydes. Pulsed amperometric detection couples the process of anodic detection with anodic cleaning and cathodic reactivation of a noble-metal electrode, thus assuring a continuously cleaned and active surface. This is usually accomplished with a three-step potential waveform, combining anodic and cathodic polarizations (e.g., see Fig. 3.29). The analytical response results primarily from adsorbed analyte, with detection limits approaching 50 ng (for 50-μL samples). Other automated multistep potential waveforms are possible. Such waveforms are commonly executed at a frequency of 1–2 Hz, in connection with gold or platinum working electrodes.

The power of electrochemical detection can be improved by using more than one working electrode (89). Different strategies, based primarily on dual-electrode detection, can thus be employed. For example, in the series mode (Fig. 3.30, top) the first upstream electrode can be used to generate an electroactive species that is then more easily detected at the downstream electrode. Discrimination against compounds with irreversible redox chemistry can also be achieved. Significantly improved qualitative information can be

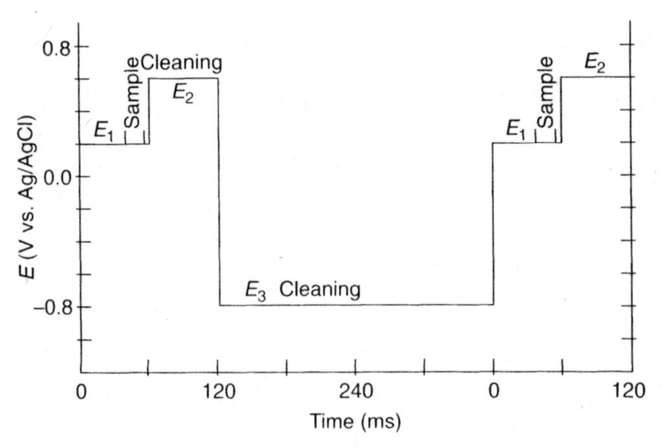

Figure 3.29 Triple-pulse amperometric waveform.

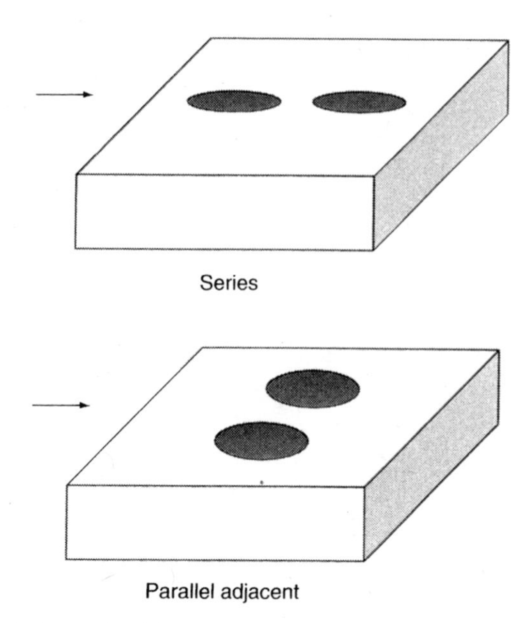

Figure 3.30 Dual-electrode thin-layer detector configurations for operation in the series (top) and parallel (bottom) amperometric modes.

achieved using a parallel (side-by-side) dual-electrode configuration (Fig. 3.30, bottom). Two simultaneous chromatograms can be generated by holding these electrodes at different potentials. The current ratios at these two potential settings provide real-time "fingerprints" of the eluting peaks. Such ratio values are compared with those of standards to confirm the peak identity. Further improvements in the information content can be achieved using multichannel amperometric detection (analogous to diode array optical detection) (90). For example, Figure 3.31 displays a three-dimensional chromatogram for a mixture of several biologically significant compounds at a 16-electrode detector array. By rapidly applying a five-potential sequence to the individual electrodes, an 80-channel chromatographic detection can be obtained. Such an electrochemical profile across the array provides confirmation of peak purity and improved identification of target analytes. Additional information can be obtained by using arrays comprised of different electrode materials (see Section 6.4). For a more detailed description of on-line electrochemical detectors, the reader is referred to a monograph by Stulik and Pacáková (91). Comparison of various commercial detectors is also available (92).

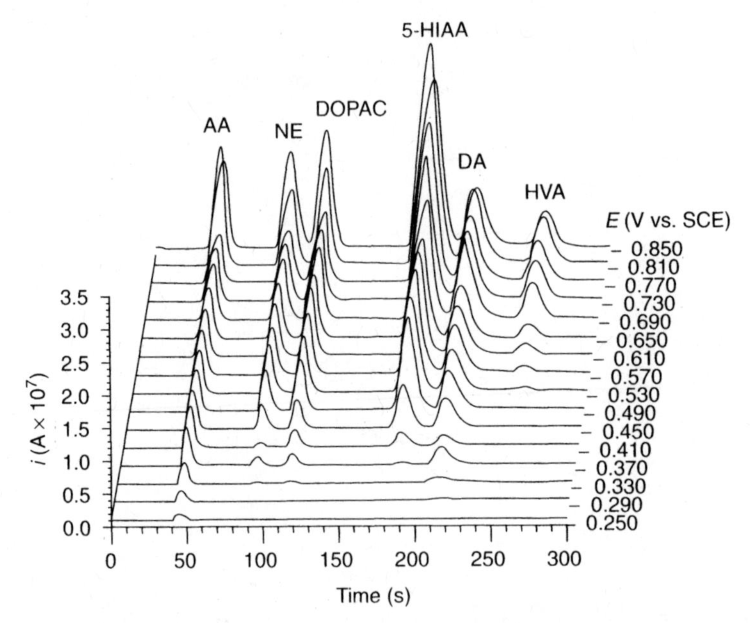

Figure 3.31 Three-dimensional chromatogram for oxidizable biological compounds at a multichannel amperometric detection system, consisting of an array of 16 carbon paste electrodes poised at different potentials. (Reproduced with permission from Ref. 90.)

EXAMPLES

Example 3.1 Voltammogram *a* was obtained for adsorptive stripping measurements of Fe(III) in seawater. Voltammograms *b* and *c* show successive standard additions of 4 ppb Fe(III). Find the concentration of Fe(III) in the sample.

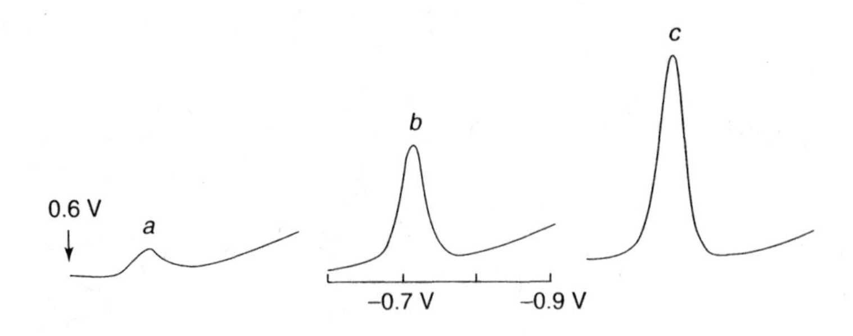

Solution The resulting current peaks lead to the following standard addition plot:

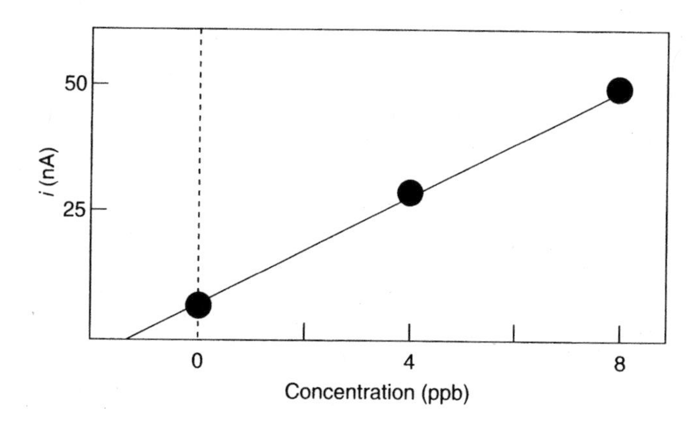

From this plot, an Fe(III) concentration of 1.47 ppb can be obtained for the sample.

Example 3.2 Calculate the limiting current that would be expected from the reduction of 2×10^{-4} M Pb^{2+}, using the DME characteristics, $m = 2.0$ mg/s and $t = 4$ s. The diffusion coefficient of Pb^{2+} is 1.01×10^{-5} cm^2/s.

Solution The lead reduction is a two-reduction process:

$$Pb^{2+} + 2e^- \rightleftharpoons Pb$$

Hence, from the Ilkovic equation, (2.4), we obtain

$$I_d = 708 \times 2 \times (1.01 \times 10^{-5})^{1/2} (2.0)^{2/3} \times 4^{1/6} \times 0.2 = 1.81\,\mu A$$

Example 3.3 Draw the DC polarographic response for a mixture containing 3 mM Cu^{2+}, 2 mM Zn^{2+}, and 1 mM Cd^{2+}. The half-wave potentials for the Cu, Zn, and Cd ions are −0.12, −0.95, and −0.62 V, respectively.

Solution

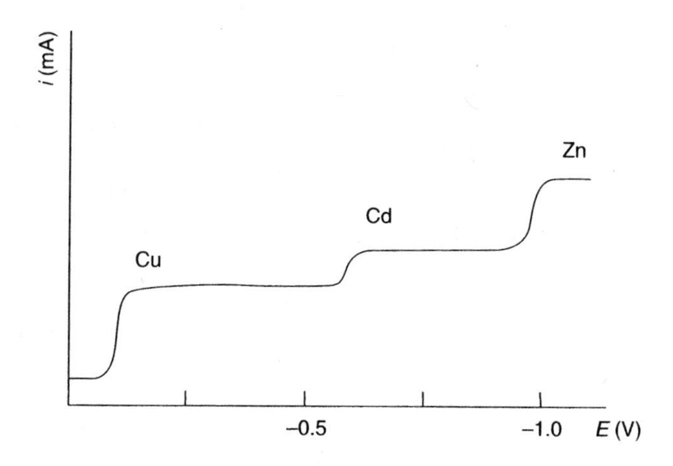

Example 3.4 A sample containing cadmium gives a polarographic reduction current of 6.0μA. The current increases to 9 and 12μA when the cadmium concentration is increased in two steps of 2mM each. Calculate the cadmium concentration in the original sample.

Solution

$$6 = KC$$
$$9 = K(C + 2)$$
$$12 = K(C + 4)$$

to yield a C value of 4mM.

Example 3.5 Polarogram *a* was obtained for a 10-mL lead-containing sample. The limiting current increased (to *B*) after addition of 100μL of a 0.10-M lead standard to the 10-mL sample. Calculate the original lead concentration in the sample.

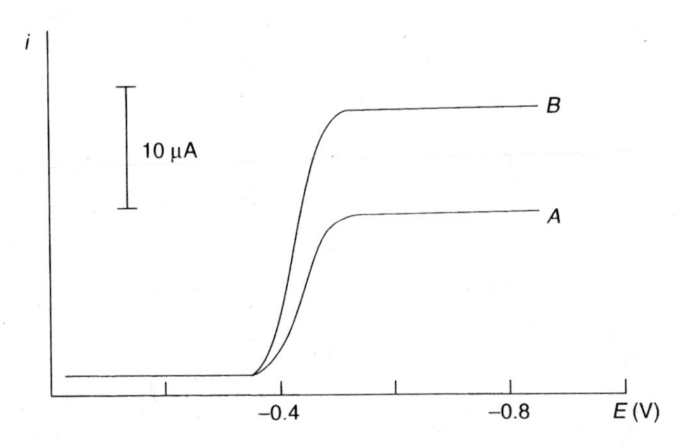

Solution The sample lead ion yielded a limiting current of 13μA (*A*). The current increases by 8.5μA on spiking the sample with a 1-mM lead standard (considering the 1 : 100 fold dilution; *B*-*A*).

$$i_l = KC \qquad 8.5 = K(1\,\text{mM})$$
$$K = 8.5$$
$$13.0 = 8.5C_{\text{sample}} \qquad C_{\text{sample}} = 1.53\,\text{mM}$$

Example 3.6 Flow analysis of a urine sample at a thin-layer amperometric detector, with a flow-rate of 1.25mL/min, yielded a limiting current value of 1.6μA for its unknown uric acid content. A larger current of 2.4μA was observed for a sample containing 1×10^{-4}M uric acid and flowing at a rate of 0.9mL/min. Calculate the original concentration of uric acid in the sample.

Solution From Table 3.4, we obtain

$$i_l = KCU^{1/3}$$

$$2.4 = K(1 \times 10^{-4})0.9^{1/3} \qquad K = 2.49 \times 10^4$$

$$1.6 = 2.49 \times 10^4 (C)1.25^{1/3} \qquad C = 6 \times 10^{-5}\ \mathrm{M}$$

PROBLEMS

3.1 Describe the principle and operation of potentiometric stripping analysis (PSA). How does it differ from anodic stripping voltammetry (ASV)? What is the quantitative signal? What, if any, are its advantages over ASV?

3.2 Draw schematic diagrams of a thin-layer flow detector utilizing an (a) single working electrode and (b) a dual electrode. Explain how the latter improves the power and information content.

3.3 The detection of nitroaromatic explosives in seawater requires a fast (1-s) and sensitive response (down to the 10 nM level). Discuss an electrochemical technique most suitable for such assays and the optimization of its variables for achieving this important goal. Clarify your choice. What is the basis for the observed response? What are the potential interferences?

3.4 Describe and draw the waveform employed in square-wave voltammetry. Explain how the current is being measured.

3.5 Why is it essential to wait ~40 ms after the potential step in normal pulse polarography before sampling the current?

3.6 Describe the use of polarographic analysis for obtaining the values of the formation constant and stoichiometric number of metal complexes.

3.7 Describe the challenges of interfacing electrochemical detectors to capillary electrophoresis separation systems. How can these challenges be addressed?

3.8 Explain how and why the coupling of stripping voltammetry to a differential-pulse waveform can enhance the power of stripping measurements of trace metals.

3.9 Select an electrochemical technique most suitable for detecting trace levels of nickel in groundwater. Explain your choice.

3.10 A liquid chromatographic experiment resulted in the same retention time for the electroactive compounds A and B ($E°$A = +0.43 V; $E°$ B = +0.77 V). Which electrochemical detection scheme would offer a selective detection of the two coeluting analytes? Explain your selection.

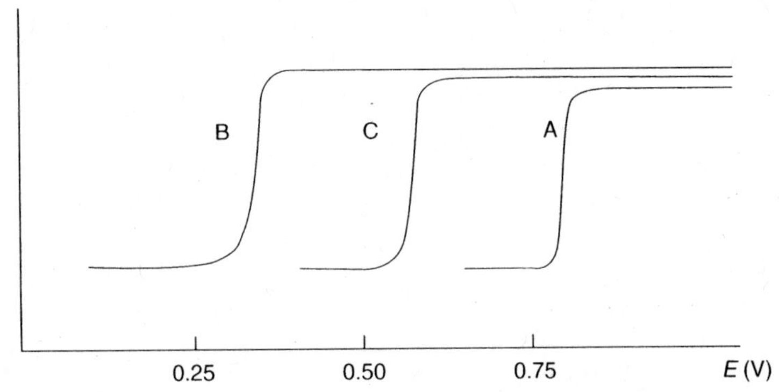

3.11 Use the adsorption theory (of Section 2.1.3) to explain why adsorptive stripping voltammetry results in nonlinear calibration plots.

3.12 Derive the Cottrell equation by combining Fick's first law of diffusion with the time-dependent change of the concentration gradient during a potential step experiment.

3.13 The oxidizable compounds A, B, and C display the following hydrodynamic voltammograms:

Determine which operating potential will allow (a) selective flow injection detection of B in the presence of A and C and (b). selective HPLC measurements of A ($t_{r,A} = 18$ min; $t_{r,B} = 18$ min, $t_{r,C} = 11$ min); explain your choices.

3.14 While carbohydrates and alcohols can be oxidized at gold electrodes, they cannot be detected by fixed-potential amperometry. Explain why, and suggest an alternative more suitable detection scheme for their measurements in flowing streams.

REFERENCES

1. Ilkovic, D., *Coll. Czech. Chem. Commun.* **64**, 498 (1934).
2. Lingane, J., *Chem. Rev.* **29**, 1 (1941).
3. Barker, G. C.; Jenkin, I. L., *Analyst* **77**, 685 (1952).
4. Barker, G. C.; Gardner, A. W., *Z. Anal. Chem.* **173**, 79 (1960).
5. Osteryoung, J.; Kirowa-Eisner, E., *Anal. Chem.* **52**, 62 (1980).
6. Parry, E. P.; Osteryoung, R. A., *Anal. Chem.* **37**, 1634 (1964).
7. Flato, J. B., *Anal. Chem.* **44**, 75A (1972).
8. Osteryoung, J.; Osteryoung, R. A., *Anal. Chem.* **57**, 101A (1985).
9. O'Dea, J. J.; Osteryoung, J.; Osteryoung, R. A., *Anal. Chem.* **53**, 695 (1981).
10. Borman, S., *Anal. Chem.* **54**, 698A (1982).
11. Wang, J.; Lu, F.; MacDonald, D.; Lu, J.; Ozsoz, M.; Rogers, K., *Talanta* **46**, 1405 (1998).
12. Yarnitzky, C., *Anal. Chem.* 57, 2011 (1985).
13. Wang, J.; Ouziel, E.; Yarnitzky, C.; Ariel, M., *Anal. Chim. Acta* **102**, 99 (1978).

14. Samuelsson, R.; O'Dea, J. J.; Osteryoung, J., *Anal. Chem.* **52**, 2215 (1980).
15. Gerhardt, G. C.; Cassidy, R. M.; Baranski, A. S., *Anal. Chem.* **70**, 2167 (1998).
16. Stefani, S.; Seeber, R., *Anal. Chem.* **54**, 2524 (1982).
17. Breyer, B.; Bauer, H., *Rev. Polarogr.* **8**, 157 (1960).
18. Smith, D. E., *CRC Crit. Rev. Anal. Chem.* **2**, 247 (1971).
19. Breyer, B.; Bauer, H., *Alternating Current Polarography and Tensammetry*, Wiley-Interscience, New York, 1963.
20. Brazill, S. A.; Bender, S. E.; Hebert, N. E.; Cullison, J. K.; Kristensen, E. W.; Kuhr, W. G., *J. Electroanal. Chem.* **531**, 119 (2002).
21. Hebert, N.; Kuhr, W.; Brazill, S., *Anal. Chem.* **75**, 3301 (2003).
22. Zhang, J.; Guo, S.; Bond, A. M.; Marken, F., *Anal. Chem.* **76**, 3619 (2004).
23. Wang, J., *Stripping Analysis: Principles, Instrumentation and Applications*, VCH Publishers, Deerfield Beach, FL, 1985.
24. Copeland, T. R.; Skogerboe, R. K., *Anal. Chem.* **46**, 1257A (1974).
25. Florence, T. M., *J. Electroanal. Chem.* **27**, 273 (1970).
26. Economou, A.; Fielden, P. R., *Analyst* **128**, 205 (2003).
27. Wang, J.; Lu, J.; Hocevar, S.; Farias, P.; Ogorevc, B., *Anal. Chem.* **72**, 3218 (2000).
28. Wang, J., *Electroanalysis* **17**, 1341 (2005).
29. Marken, F.; Rebbitt, T. O.; Booth, J.; Compton, R. G., *Electroanalysis* **9**, 19 (1997).
30. Jagner, D., *Trends Anal. Chem.* **2**(3), 53 (1983).
31. Wang, J., "Voltammetry after nonelectrolytic proconcentration," in A. J. Bard, ed., *Electroanalytical Chemistry*, Marcel Dekker, New York, 1989, Vol. 16, p. 1.
32. Van den Berg, C. M. G., *Anal. Chim. Acta* **250**, 265 (1991).
33. Paneli, M.; Voulgaropoulos, A., *Electroanalysis* **5**, 355 (1993).
34 Van den Berg, C. M. G.; Murphy, K.; Riley, J. P., *Anal. Chim. Acta* **188**, 177 (1986).
35. Wang, J.; Baomin, T., *Anal. Chim. Acta* **270**, 137 (1992).
36. Donat, J. R.; Bruland, K. W., *Anal. Chem.* **60**, 240 (1988).
37. Golimowski, J.; Valenta, P.; Nürnberg, H. W., *Fres. Z. Anal. Chem.* **322**, 315(1985).
38. Wang, J.; Mahmoud, J. S., *Fres. Z. Anal. Chem.* **327**, 789 (1987).
39. Wang, J.; Mahmoud, J. S., *J. Electroanal. Chem.* **208**, 383 (1986).
40. van den Berg, C. M. C., *Anal. Chem.* **57**, 1532 (1985).
41. Pihlar, B.; Valenta, P.; Nürnberg, H. W., *Fres. Z. Anal. Chem.* **307**, 337 (1981).
42. Wang, J.; Zadeii, J.; Lin, M. S., *J. Electroanal. Chem.* **237**, 281 (1987).
43. Wang, J.; Zadeii, J., *Talanta* **34**, 909 (1987).
44. Yokoi, K.; van den Berg, C. M. C., *Anal. Chim. Acta* **245**, 167 (1991).
45. Van den Berg, C. M. G.; Nimmo, N., *Anal. Chem.* **59**, 269 (1987).
46. Van den Berg, C. M. G., *Anal. Chem.* **56**, 2383 (1984).
47. Wang, J.; Cai, X.; Jonsson, C.; Balakrishan, M., *Electroanalysis* **8**, 20 (1996).
48. Luther, G. W.; Swartz, C.; Ullman, W., *Anal. Chem.* **60**, 1721 (1988).
49. Scholz, F.; Lange, B., *Trends Anal. Chem.* **11**, 359 (1992).
50. Tercier, M.; Buffle, J., *Electroanalysis* **5**, 187 (1993).
51. Komanur, N. K.; van Loon, G. W., *Talanta* **24**, 184 (1977).
52. Poldoski, J.; Glass, G., *Anal. Chim. Acta* **101**, 79 (1978).

53. Wang, J.; Lu, J.; Olsen, K., *Analyst* **117**, 1913 (1992).
54. Hutton, E.; van Elteren, J.; Ogorevc, B.; Smyth, M., *Talanta* **63**, 849 (2004).
55. Gottesfeld, S.; Ariel, M., *J. Electroanal. Chem.* **9**, 112 (1965).
56. Wang, J.; Mannino, S., *Analyst* **114**, 643 (1989).
57. Obata, H.; van den Berg C. M. C., *Anal. Chem.* **73**, 2522 (2001).
58. Ostapczuk, P., *Clin. Chem.* **38**, 1995 (1992).
59. Lai, P.; Fung, K., *Analyst* **103**, 1244 (1978).
60. Golimowski, J.; Gustavsson, I., *Fres. Z. Anal. Chem.* **317**, 484 (1984).
61. Adeloju, S. B.; Bond, A. M.; Briggs, M. H., *Anal. Chim. Acta* **164**, 181 (1984).
62. Hoppstock, K.; Michulitz, M., *Anal. Chim. Acta* **350**, 135 (1997).
63. Porbes, S.; Bound, G.; West, T., *Talanta* **26**, 473 (1979).
64. Levit, D. I., *Anal. Chem.* **45**, 1291 (1973).
65. Wang, J.; Setiadji, R.; Chen, L.; Lu, J.; Morton, S., *Electroanalysis* **4**, 161 (1992).
66. Williams, T.; Foy, O.; Benson, C., *Anal. Chim. Acta* **75**, 250 (1975).
67. Cavicchioli, A.; La-Scalea, M. A.; Gutz, I. G., *Electroanalysis* **16**, 697 (2004).
68. Florence, T. M., *Analyst* **111**, 489 (1986).
69. Wang, J., *Analyst* **119**, 763 (1994).
70. Tercier, M. L.; Buffle, J.; Graziottin, F., *Electroanalysis* **10**, 355 (1998).
71. Wang, J., *Anal. Chim. Acta* **500**, 247 (2003).
72. Moane, S.; Park, S.; Lunte, C. E.; Smyth, M. R., *Analyst* **123**, 1931 (1998).
73. Ewing, A. G.; Mesaros, J. M.; Gavin, P. F., *Anal. Chem.* **66**, 527A (1994).
74. Curry, P.; Engstrom, C.; Ewing, A., *Electroanalysis*, **3**, 587 (1991).
75. Holland, L. A.; Lunte, S. M., *Anal. Commun.* **35**, 1H (1998).
76. Sloss, S.; Ewing, A. G., *Anal. Chem.* **63**, 577 (1993).
77. Woolley, A.; Lao, K.; Glazer, A.; Mathies, R., *Anal. Chem.* **70**, 684 (1998).
78. Wang, J., *Electroanalysis* **17**, 1133 (2005).
79. Lacher, N. A.; Garrison, K. E.; Martin, R. S.; Lunte, S. M., *Electrophoresis* **22**, 2526 (2001).
80. Woolley, T.; Lao, K.; Glazer, A. N.; Mathies, R. A., *Anal. Chem.* **70**, 684 (1998).
81. Wang, J.; Tian, B.; Sahlin, E., *Anal. Chem.* **71**, 5436 (1999).
82. Hilmi, A.; Luong, J. H., *Anal. Chem.* **72**, 4677 (2000).
83. Chen, D.; Hsu, F.; Zhan, D.; Chen, C., *Anal. Chem.* **73**, 758 (2001).
84. Hanekamp, H. B.; Box, P.; Frei, R. W., *Trends Anal. Chem.* **1**, 135 (1982).
85. Trojanek, A.; De Jong, H. G., *Anal. Chim. Acta* **141**, 115 (1982).
86. Johnson, D. C.; LaCourse, W. R., *Anal. Chem.* **62**, 589A (1990).
87. Johnson, D.C.; LaCourse, W. R., *Electroanalysis* **4**, 367 (1992).
88. LaCourse, W. R., *Pulsed Electrochemical Detection in HPLC*, Wiley, New York, 1997.
89. Roston, D. A.; Shoup, R. E.; Kissinger, P. T., *Anal. Chem.* **54**, 1417A (1982).
90. Hoogvliet, J.; Reijn, J.; van Bennekom, W., *Anal. Chem.* **63**, 2418 (1991).
91. Stulik, K.; Pacáková, V., *Electroanalytical Measurements in Flowing Liquids*, Ellis Horwood, Chichester, UK, 1987.
92. Warner, M., *Anal. Chem.* **66**, 601A (1994).

4

PRACTICAL CONSIDERATIONS

The basic instrumentation required for controlled-potential experiments is relatively inexpensive and readily available commercially. The basic necessities include a cell (with a three-electrode system), a voltammetric analyzer (consisting of a potentiostatic circuitry and a voltage ramp generator), and a plotter. Modern voltammetric analyzers are versatile enough to perform many modes of operation. Depending on the specific experiment, other components may be required. For example, a faradaic cage is desired for work with ultramicroelectrodes. The system should be located in a room free from major electrical interferences, vibrations, and drastic fluctuations in temperature.

4.1 ELECTROCHEMICAL CELLS

Three-electrode cells (e.g., see Fig. 4.1) are commonly used in controlled-potential experiments. The cell is usually a covered beaker of 5–50 mL volume, and contains the three electrodes (working, reference, and auxiliary), which are immersed in the sample solution. While the working electrode is the electrode at which the reaction of interest occurs, the reference electrode provides a stable and reproducible potential (independent of the sample composition), against which the potential of the working electrode is compared. Such "buffering" against potential changes is achieved by a constant composition of both forms of its redox couple, such as Ag/AgCl or Hg/Hg_2Cl_2, as common

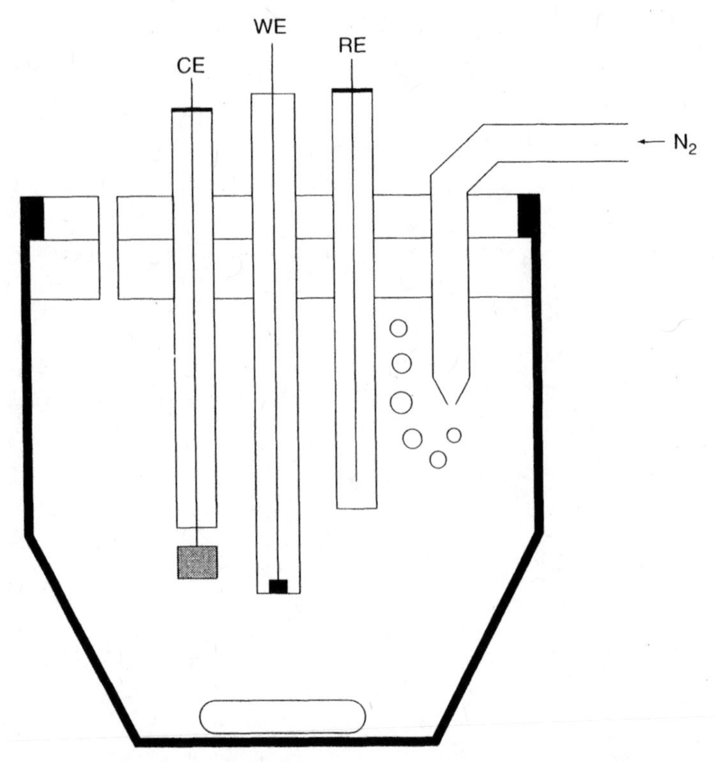

Figure 4.1 Schematic diagram of a cell for voltammetric measurements: WE—working electrodes; RE—reference electrode; CE—counter electrode. The electrodes are inserted through holes in the cell cover.

with the silver–silver chloride and the saturated calomel reference electrodes, respectively. To minimize contamination of the sample solution, the reference electrode may be insulated from the sample through an intermediate bridge. An inert conducting material, such as platinum wire or graphite rod, is usually used as the current-carrying auxiliary electrode. The relative position of these electrodes and their proper connection to the electrochemical analyzer should be noted (see Section 4.4). The three electrodes, as well as the tube used for bubbling the deoxygenating gas (see Section 4.3), are supported in five holes in the cell cover. Complete systems, integrating the three-electrode cell, built-in gas control, and magnetic stirrer, along with proper cover, are available commercially (e.g., see Fig. 4.2).

The exact cell design and the material used for its construction are selected according to the experiment at hand and the nature of the sample. The various designs differ with respect to size, temperature control capability, stirring requirement, shape, or number of cell compartments. Various microcells with 20–500 μL volumes can be used when the sample volume is limited. Particularly attractive are thin-layer cells in which the entire sample is confined within

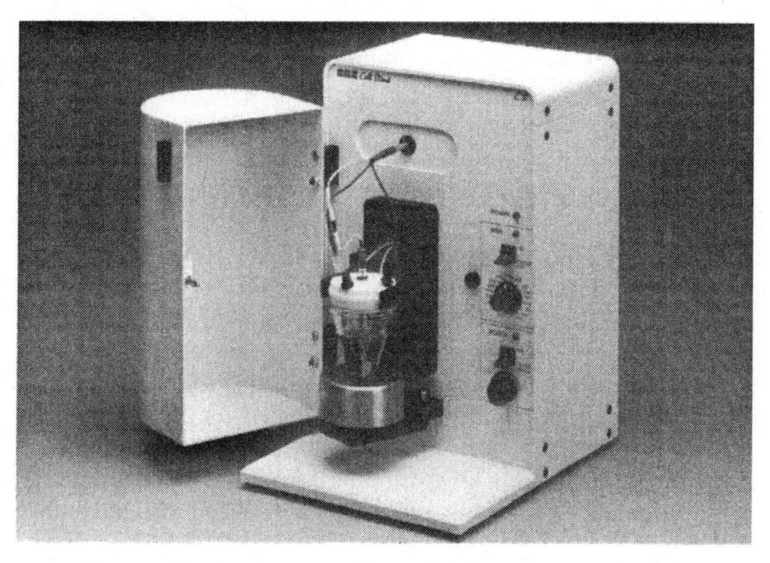

Figure 4.2 A complete cell stand. (Courtesy of BAS Inc.)

a thin layer (of less than 10 μm thickness) at the electrode surface (1). Smaller sample volumes can be accommodated in connection with ultramicroelectrodes (discussed in Section 4.4) and advanced microfabrication processes (discussed in Section 6.3). In particular, lithographically fabricated picoliter microvials (2) hold great promise for assays of ultrasmall environments (e.g., single-cell systems). Specially designed flow cells (discussed in Section 3.6) are used for on-line applications. Glass is commonly used as the cell material, due to its low cost, transparency, chemical inertness, and impermeability. Teflon and quartz represent other possible cell materials. The cell cover can be constructed of any suitable material that is inert to the sample. An accurate temperature control is readily achieved by immersing or jacketing the cell in a constant-temperature bath.

4.2 SOLVENTS AND SUPPORTING ELECTROLYTES

Electrochemical measurements are commonly carried out in a medium that consists of solvent containing a supporting electrolyte. The choice of solvent is dictated primarily by the solubility of the analyte and its redox activity, and by solvent properties, such as the electrical conductivity, electrochemical activity, and chemical reactivity. The solvent should not react with the analyte (or products) and should not undergo electrochemical reactions over a wide potential range.

While water has been used as a solvent more than any other medium, nonaqueous solvents [e.g., acetonitrile, propylene carbonate, dimethylformamide

(DMF), dimethylsulfoxide (DMSO), or methanol] have also frequently been used. Mixed solvents may also be considered for certain applications. Double-distilled water is adequate for most work in aqueous media. Triple-distilled water is often required when trace (stripping) analysis is concerned. Organic solvents often require a drying or purification procedure. These and other solvent-related considerations have been reviewed by Mann (3).

Supporting electrolytes are required in controlled-potential experiments to decrease the resistance of the solution, eliminate electromigration effects, and maintain a constant ionic strength (i.e., "swamping out" the effect of variable amounts of naturally occurring electrolyte) (4). The inert supporting electrolyte may be an inorganic salt, a mineral acid, or a buffer. While potassium chloride or nitrate, ammonium chloride, sodium hydroxide, or hydrochloric acid are widely used when water is employed as a solvent, tetraalkylammonium salts are often employed in organic media. Buffer systems (such as acetate, phosphate, or citrate) are used when a pH control is essential. The composition of the electrolyte may affect the selectivity of voltammetric measurements. For example, the tendency of most electrolytes to complex metal ions can benefit the analysis of mixtures of metals. In addition, masking agents [such as ethylenediaminetetraacetic acid (EDTA)] may be added to "remove" undesired interferences. The supporting electrolyte should be prepared from highly purified reagents, and should not be easily oxidized or reduced (hence minimizing potential contamination or background contributions, respectively). The usual electrolyte concentration range is 0.1–1.0 M, i.e., in large excess of the concentration of all electroactive species. Significantly lower levels can be employed in connection with ultramicroscale working electrodes (see Section 4.5.4).

4.3 OXYGEN REMOVAL

The electrochemical reduction of oxygen usually proceeds via two well-separated two-electron steps. The first step corresponds to the formation of hydrogen peroxide

$$O_2 + 2H^+ + 2e^- \rightarrow H_2O_2 \qquad (4.1)$$

and the second step corresponds to the reduction of the peroxide:

$$H_2O_2 + 2H^+ + 2e^- \rightarrow 2H_2O \qquad (4.2)$$

The half-wave potentials of these steps are approximately −0.1 and −0.9 V (vs. the saturated calomel electrode). The exact stoichiometry of these steps is dependent on the medium. The large background current accrued from this stepwise oxygen reduction interferes with the measurement of many reducible analytes. In addition, the products of the oxygen reduction may affect the electrochemical process under investigation.

A variety of methods have thus been used for the removal of dissolved oxygen (5). The most common method has been purging with an inert gas (usually purified nitrogen) for 4–8 min prior to recording of the voltammogram. Longer purge times may be required for large sample volumes or for trace measurements. To prevent oxygen from reentering, the cell should be blanketed with the gas while the voltammogram is being recorded. Passage of the gas through a water-containing presaturator is desired to avoid evaporation. The deaeration step, although time-consuming, is quite effective and suitable for batch analysis. (The only exception is work with microsamples, where deoxygenation may lead to errors caused by the evaporation of solvent or loss of volatile compounds.)

Other methods have been developed for the removal of oxygen (particularly from flowing streams). These include the use of electrochemical or chemical (zinc) scrubbers, nitrogen-activated nebulizers, and chemical reduction (by addition of sodium sulfite or ascorbic acid). Alternately, it may be useful to employ voltammetric methods that are less prone to oxygen interference. The background-correction capability of modern (computerized) instruments is also effective for work in the presence of dissolved oxygen.

4.4 INSTRUMENTATION

Rapid advances in microelectronics, and in particular the introduction of operational amplifiers, have led to major changes in electroanalytical instrumentation. Tiny and inexpensive integrated circuits can now perform many functions that previously required very large instruments. Such trends have been reviewed (6). Various voltammetric analyzers are now available commercially from different sources (Table 4.1) at relatively modest prices [ranging from $5000 to $25,000 (in 2005)]. Such instruments consist of two circuits: a polarizing circuit that applies the potential to the cell, and a measuring circuit that monitors the cell current. The characteristic of modern voltammetric analyzers is the potentiostatic control of the working electrode, which minimizes errors due to cell resistance (i.e., poorly defined voltammograms with lower current response and shifted and broadened peaks). Equation 4.3 explains the cause for this ohmic distortion:

$$E_{app} = E_{WE} - E_{RE} - iR \tag{4.3}$$

where iR is the ohmic potential drop.

The potentiostatic control, aimed at compensating a major fraction of the cell resistance, is accomplished with a three-electrode system and a combination of operational amplifiers and feedback loops (Fig. 4.3). Here, the reference electrode is placed as close as possible to the working electrode and

TABLE 4.1 Current Suppliers of Voltammetric Analyzers

Supplier	Address
Analytical Instrument Systems	PO Box 458 Flemington, NJ 08822 www.aishome.com
Bioanalytical Systems	2701 Kent Ave. W. Lafayette, IN 47906 www.bioanalytical.com
Cypress	PO Box 3931 Lawrence, KS 66044 www.cypresshome.com
CH Instruments	3700 Tennison Hill Dr. Austin, TX 78733 chinstr@worldnet.att.net
ECO Chemie	PO Box 85163 3508 AD Utrecht The Netherlands autolab@ecochemie.nl www.brinkmann.com
EG&G PAR	801 S. Illinois Ave. Oak Ridge, TN 37830 www.egg.inc.com/par
ESA	45 Wiggins Ave. Bedford, MA 01730
Metrohm	CH-9109 Herisau Switzerland www.brinkmann.com
Palm Instruments BV	Ruitercamp 119 3993 BZ Houten The Netherlands www.palmsens.com
Radiometer/Tacussel	27 rue d'Alscace F-69627 Villeurbanne France Analytical@clevelandOH.com
Solartron	964 Marcon Blvd. Allentown, PA 18103, USA www.solartron.com
TraceDetect	Seattle, WA, USA www.tracedetect.com

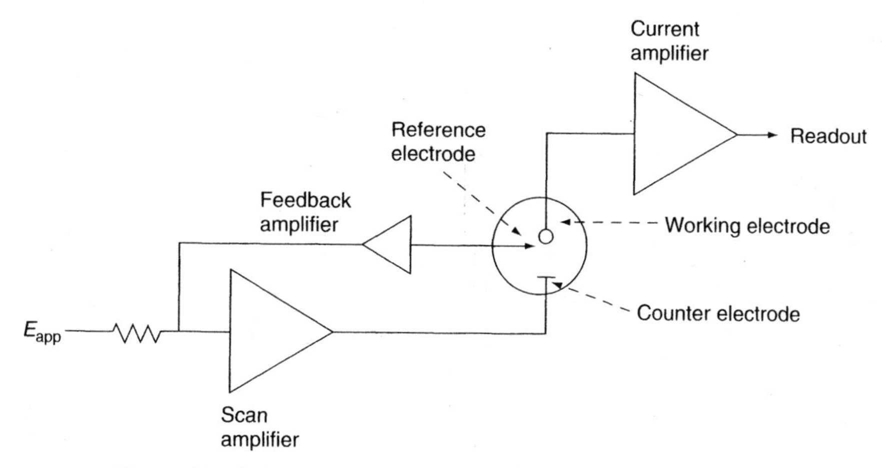

Figure 4.3 Schematic diagram of a three-electrode potentiostat.

connected to the instrument through a high-resistance circuit that draws no current from it. Because the flow cannot occur through the reference electrode, a current-carrying auxiliary electrode is placed in the solution to complete the current path. Hence, the current flows through the solution between the working and the auxiliary electrodes. Symmetry in the placement of these electrodes is important for the assumption that the current paths from all points on the working electrode are equivalent. Because no current passes through the reference electrode and because of its position close to the working electrode, the potential drop caused by the cell resistance (iR) is minimized. If the potential sensed by the reference electrode is less than the desired value, the operational amplifier control loop provides a corrective potential. By adding an operational amplifier current-to-voltage converter (called a "current follower") to the working electrode, it is possible to measure the current without disturbing the controlled parameters. The instrument also includes a ramp generator to produce various regularly changing potential waveforms.

As was pointed out earlier, an effective potential control requires a very close proximity between the working and reference electrodes. This can be accomplished by using a specially designed bridge of the reference electrode, known as a *Luggin probe*. The tip of this bridge should be placed as close as twice its diameter to the working electrode. A smaller distance will result in blockage (shielding) of the current path and hence a nonhomogeneous current density. The Luggin bridge should also not interfere to the convective transport toward the surface of the working electrode.

It should be pointed out that not all of the iR drop is removed by the potentiostatic control. Some fraction, denoted as iR_u (where R_u is the uncompensated solution resistance between the reference and working electrodes), will still be included in the measured potential. This component may be significantly large when resistive nonaqueous media are used, and thus may lead to severe distortion of the voltammetric reponse. Many modern instruments, however, automatically subtract (compensate) the iR_u drop from the potential signal given to the potentiostat via an appropriate positive feedback.

Biopotentiostats, offering simultaneous control of two working electrodes (e.g., in ring–disk configuration) are also available. Such instruments consist of a conventional potentiostat with a second voltage-control circuit. Multipotentiostats, controlling multiple working electrodes (connected to a multiplexed data acquisition circuitry), have also been described (7). The development of ultramicroelectrodes, with their very small currents (and thus negligible iR losses even when R is large), allows the use of simplified, two-electrode, potential control (see Section 4.5.4). In contrast, ultramicroelectrode work requires an efficient current measurement circuitry to differentiate between the faradaic response and the extraneous electronic noise and for handling low currents down to the pA (picoampere) range. Other considerations for noise reduction involve the grounding and shielding of the instrument and cell.

The advent of inexpensive computing power has changed dramatically the way voltammetric measurements are controlled and data are acquired and manipulated. Computer-controlled instruments, available from most manufacturers (6), provide flexibility and sophistication in the execution of a great variety of modes. In principle, any potential waveform that can be defined mathematically can be applied with commands given through a keyboard. Such instruments offer various data processing options, including autoranging, blank subtraction, noise reduction, curve smoothing, differentiation, integration, and peak search. The entire voltammogram can be presented as a plot or printout (of the current–potential values). In addition, computer control has allowed automation of voltammetric experiments and hence has greatly improved the speed and precision of the measurement. Since the electrochemical cell is an analog element, and computers work only in the digital domain, analog-to-digital (A/D) and digital-to-analog (D/A) converters are used to interface between the two. Unattended operation has been accomplished through the coupling of autosamplers and microprocessor-controlled instruments (e.g., see Fig. 4.4). The autosampler can accommodate over 100 samples, as well as relevant standard solutions. Such coupling can also address the preliminary stages of sample preparation (as dictated by the nature of the sample). The role of computers in electroanalytical measurements and in the development of "smarter" analyzers has been reviewed by Bond (8) and He et al. (9).

The nature of electrochemical instruments makes them very attractive for decentralized testing. For example, compact, battery-operated voltammetric analyzers, developed for on-site measurements of metals (e.g., 10,11), readily address the growing needs for field-based environmental studies and security

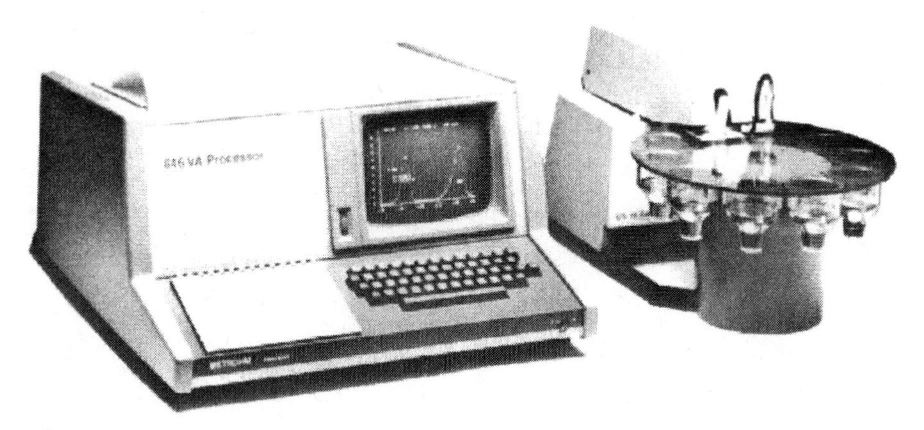

Figure 4.4 Microprocessor-controlled voltammetric analyzer, in connection with an autosampler. (Courtesy of Metrohm Inc.)

surveillance applications. Similarly, portable (hand-held) instruments are being designed for decentralized clinical testing (12).

4.5 WORKING ELECTRODES

The performance of the voltammetric procedure is strongly influenced by the working-electrode material. The working electrode should provide high signal-to-noise characteristics, as well as a reproducible response. Thus, its selection depends primarily on two factors: the redox behavior of the target analyte and the background current over the potential region required for the measurement. Other considerations include the potential window, electrical conductivity, surface reproducibility, mechanical properties, cost, availability, and toxicity. A range of materials have found application as working electrodes for electroanalysis. The most popular are those involving mercury, carbon, or noble metals (particularly platinum and gold). Figure 4.5 displays the accessible potential window of these electrodes in various solutions. The geometry of these electrodes must also be considered.

4.5.1 Mercury Electrodes

Mercury is a very attractive choice for electrode materials because it has a high hydrogen overvoltage that greatly extends the cathodic potential window (compared to solid electrode materials) and possesses a highly reproducible, readily renewable, and smooth surface. In electrochemical terms, its roughness factor is equal to one (i.e., identical geometric and actual surface areas). Disadvantages of the use of mercury are its limited anodic range (due to the oxidation of mercury) and toxicity.

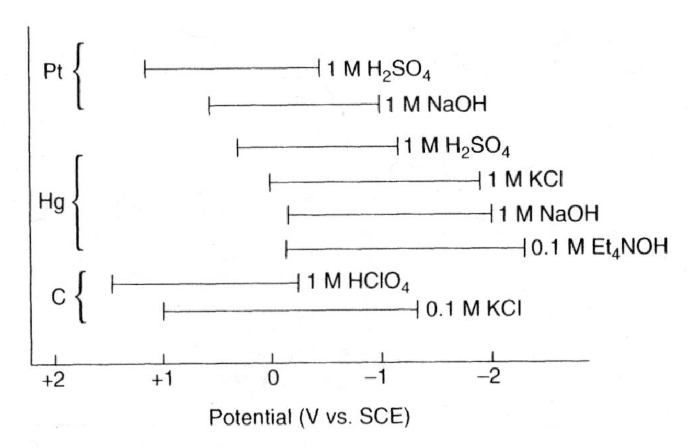

Figure 4.5 Accessible potential window of platinum, mercury, and carbon electrodes in various supporting electrolytes.

There are several types of mercury electrodes. Of these, the dropping mercury electrode (DME), the hanging mercury drop electrode (HMDE), and the mercury film electrode (MFE) are the most frequently used. Related solid amalgam electrodes have been introduced more recently to address concerns related to the toxicity of mercury.

The DME, used in polarography (Section 3.2) and for electrocapillary studies (Section 1.4), consists of a 12–20-cm-long glass capillary tubing (with an internal diameter of 30–50 μm), connected by a flexible tube to an elevated reservoir of mercury (Fig. 4.6). Electrical contact is effected through a wire inserted into the mercury reservoir. Mercury flows by gravity through the capillary at a steady rate, emerging from its tip as continuously growing drops. By adjusting the height of the mercury column, one may vary the drop time; the lifetime of the drop is typically 2–6 s. Such continuous exposure of fresh spherical drops eliminates passivation problems that may occur at stable solid electrodes. The key to successful operation of the DME is proper maintenance of its capillary (which prevents air bubbles, solution creeping, and dirt). More elaborate DMEs, based on a mechanical drop detachment at reproducible time intervals, are used for pulse polarography.

The hanging mercury drop electrode is a popular working electrode for stripping analysis and cyclic voltammetry. In this configuration, stationary mercury drops are displaced from a reservoir through a vertical capillary. Early (Kemula-type) HMDE designs rely on a mechanical extrusion (by a micrometer-driven syringe) from a reservoir through a capillary (13). The mercury reservoir should be completely filled with mercury; air must be fully eliminated. Modern HMDEs (particularly with the model 303 of EG&G PAR, shown in Fig. 4.7) employ an electronic control of the drop formation, which offers improved reproducibility and stability (14). For this purpose, a solenoid-activated valve dispenses the mercury rapidly, and the drop size is controlled

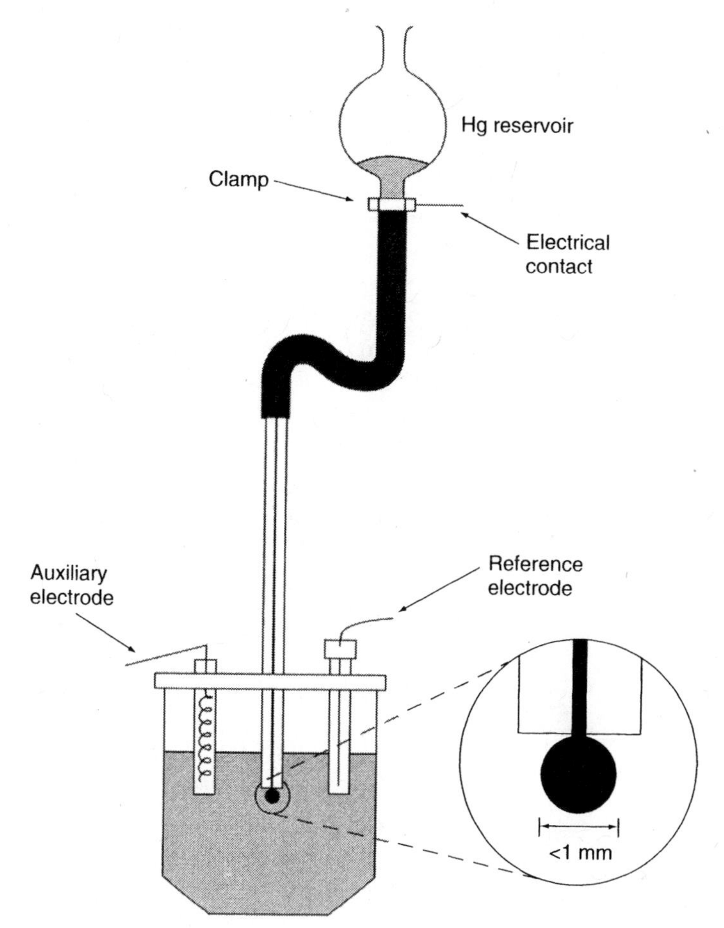

Figure 4.6 The dropping mercury electrode.

by the time during which the valve is opened. A wide-bore capillary allows a mercury drop to be grown very rapidly when the valve is opened. Three valve opening times produce drops that are described as small, medium, or large. Since the potential scan is accomplished after the valve has been closed (i.e., stationary electrode), charging-current contributions due to the drop growth are eliminated. All the components of this electrode, including the mercury reservoir, are contained in a compact unit. Such a commercial probe allows the conversion from the HMDE to DME by a single switch. When used in the DME mode, it exhibits a very rapid growth to a given area, which then remains constant (as desired for minimizing charging-current contributions). The performance of HMDEs can be improved by siliconizing the interior bore of the capillary.

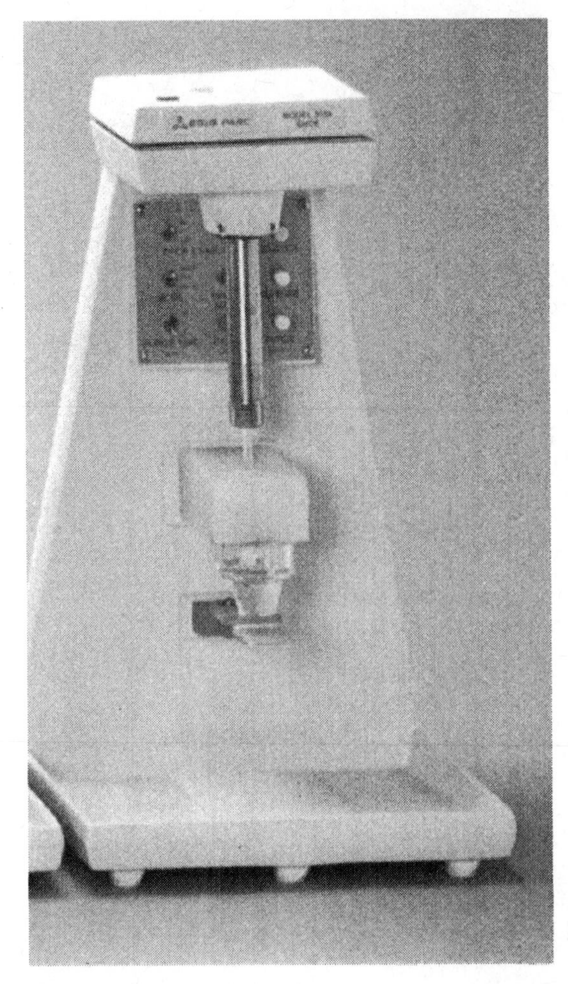

Figure 4.7 The static mercury drop electrode and its cell stand.

Several mercury electrodes combine the features of the DME and HMDE. In particular, one employs a narrow-bore capillary that produces DMEs with drop lifetimes of 50–70 s (15). The other involves a controlled growth mercury drop (16). For this purpose, a fast-response valve offers a wide range of drop sizes and a slowly (step-by-step) growing drop.

The mercury film electrode (MFE), used for stripping analysis or flow amperometry, consists of a very thin (10–100-μm) layer of mercury covering a conducting and inert support. Because of the adherent oxide films on metal surfaces, and the interaction of metals with mercury, glassy carbon is most often used as a substrate for the MFE. The mercury film formed on a glassy carbon support is actually composed of many droplets. Because they do not have a pure mercury surface, such film electrodes exhibit a lower hydrogen overvoltage and higher background currents. Another useful substrate for the MFE is iridium (because of its very low solubility in mercury and the excellent adherence of the resulting film). Mercury film electrodes are commonly preplated by cathodic deposition from a mercuric nitrate solution. An in situ plated MFE is often employed during stripping analysis (17). This electrode is prepared by simultaneous deposition of the mercury and the measured metals. Most commonly, a disk-shaped carbon electrode is used to support the mercury film. Mercury film ultramicroelectrodes, based on coverage of carbon fiber or carbon microdisk surfaces, have also received a growing attention in recent years.

4.5.2 Solid Electrodes

The limited anodic potential range of mercury electrodes has precluded their utility for monitoring oxidizable compounds. Accordingly, solid electrodes with extended anodic potential windows have attracted considerable analytical interest. Of the many different solid materials that can be used as working electrodes, the most often used are carbon, platinum, and gold. Silver, nickel, and copper can also be used for specific applications. A monograph by Adams (18) is highly recommended for a detailed description of solid electrode electrochemistry.

An important factor in using solid electrodes is the dependence of the response on the surface state of the electrode. Accordingly, the use of such electrodes requires precise electrode pretreatment and polishing to obtain reproducible results. The nature of these pretreatment steps depends on the materials involved. Mechanical polishing (to a smooth finish) and potential cycling are commonly used for metal electrodes, while various chemical, electrochemical, or thermal surface procedures are added for activating carbon-based electrodes. Unlike mercury electrodes, solid electrodes present a heterogeneous surface with respect to electrochemical activity (19). Such surface heterogeneity leads to deviations from the behavior expected for homogeneous surfaces.

Solid electrodes can be stationary or rotating, usually in a planar disk configuration. Such electrodes consist of a short cylindrical rod of the electrode material embedded in a tightly fitting tube of an insulating material (Teflon, Kel-F, etc.). The construction of a typical disk electrode is illustrated in Figure 4.8. It is essential to use proper sealing to avoid crevices between the sleeve and the electrode materials, and thus to prevent solution creeping (and an increased background response). Electrical contact is made at the rear face. Disk solid electrodes are also widely employed in flow analysis in connection with thin-layer or wall-jet detectors (see Section 3.6). Other configurations of solid electrodes, including various ultramicroelectrodes (Section 4.5.4) and microfabricated screen-printed strips or silicon-based thin-film chips (Section 6.5), are attracting increasing attention.

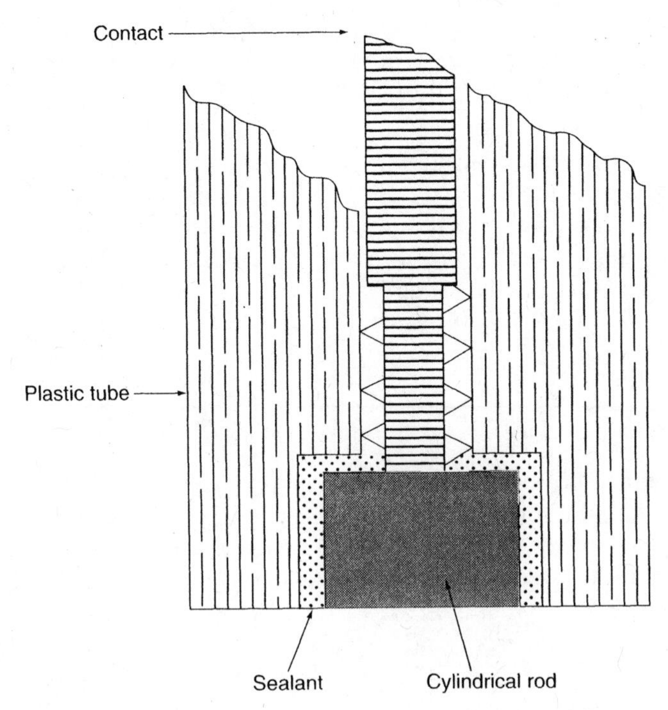

Figure 4.8 Construction of a typical disk electrode.

4.5.2.1 Rotating Disk and Rotating Ring Disk Electrodes The rotating disk electrode (RDE) is vertically mounted in the shaft of a synchronous controllable-speed motor and rotated with constant angular velocity (ω) about an axis perpendicular to the plain disk surface (Fig. 4.9a). [$\omega = 2\pi f$, where f is the rotation speed in rps (revolutions per second)]. As a result of this motion, the fluid in an adjacent layer develops a radial velocity that moves it away from the disk center. This fluid is replenished by a flow normal to the surface. Hence, the RDE can be viewed as a pump that draws a fresh solution up from the bulk solution. Under laminar flow conditions (usually up to ~4000 rpm), the thickness of the diffusion layer decreases with increasing electrode angular velocity according to

$$\delta = 1.61 D^{1/3} \omega^{-1/2} \nu^{1/6} \tag{4.4}$$

where ν is the kinematic viscosity (defined as the viscosity divided by the density in cm^2/s). Rotation speeds of 100–4000 rpm thus correspond to δ values in the 5–50 μm range. Equation (4.4) suggests that the thickness of the diffusion layer is independent of the disk diameter, namely, a uniform layer across the surface. The limiting current (for a reversible system) is thus proportional to the square root of the angular velocity, as described by the *Levich equation*:

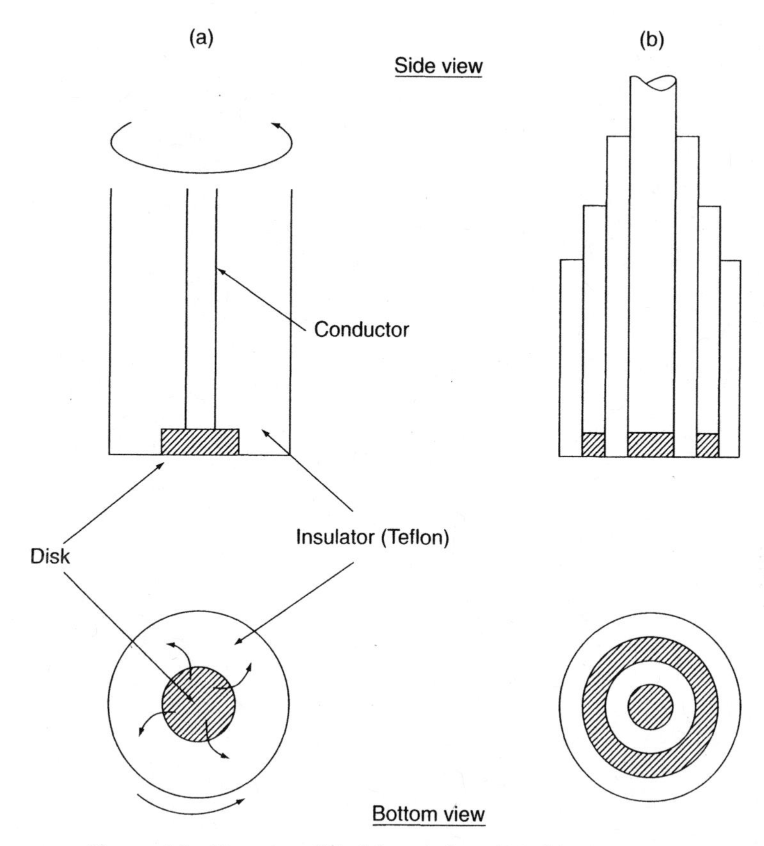

Figure 4.9 Rotating disk (a) and ring–disk (b) electrodes.

$$i_l = 0.62nFAD^{2/3}\omega^{1/2}\nu^{-1/6}C \quad (4.5)$$

An increase in ω from 400 to 1600 rpm thus results in a twofold increase of the signal. A deviation from a linear plot of i_l vs. $\omega^{1/2}$ plot suggests some kinetic limitations. In addition, at very low rotation speeds (0–100 rpm), a slight upward bend is observed, due to contribution by natural convection. The voltammetric wave has a sigmoidal shape; for reversible systems it is identical to that common in DC polarography (described in Section 3.2), and independent of ω.

For quasi-reversible systems the limiting current is controlled by both mass transport and charge transfer:

$$i_l = nFADC\left(1/1.61D^{1/3}\omega^{-1/2}\nu^{1/6} + k/D\right) \quad (4.6)$$

where k is the specific heterogeneous rate constant. In the limit of purely kinetically controlled process ($k < 10^{-6}$ m/s), the current becomes independent of the rotation speed:

$$i_l = nFAkC \tag{4.7}$$

Overall, the RDE provides an efficient and reproducible mass transport and hence analytical measurements can be made with high sensitivity and precision. Such well-defined behavior greatly simplifies interpretation of the measurement. The convective nature of the electrode results also in very short response times. The detection limits can be lowered via periodic changes in the rotation speed, and isolation of small mass-transport-dependent currents from simultaneously flowing surface-controlled background currents. Sinusoidal or square-wave modulations of the rotation speed are particularly attractive for this task. The rotation speed dependence of the limiting current [Eq. (4.5)] can also be used for calculating the diffusion coefficient or the surface area. Further details on the RDE can be found in Adams' book (18).

An extension of the RDE involves an addition of a concentric-ring electrode surrounding the disk (and separated from it by a small insulating gap) (20). The resulting rotating ring–disk electrode (RRDE), shown in Figure 4.9b, has been extremely useful for elucidating various electrode mechanisms (through generation and detection reactions at the disk and ring, respectively). Such "collection" experiments rely on measurements of the collection efficiency (N), which is the ring : disk current ratio:

$$N = -i_R / i_D \tag{4.8}$$

Here, N corresponds to the fraction of the species generated at the disk that is detected at the ring. (The negative sign arises from the fact that the currents pass in opposite directions.) Hence, the "collection" current is proportional to the "generation" current. Such experiments are particularly useful for detecting short-lived intermediate species (as the observed N values reflect the stability of such species). Analogous "collection" experiments can be carried out using dual-electrode flow detectors, described in Chapters 3–6.

4.5.2.2 Carbon Electrodes Solid electrodes based on carbon are currently in widespread use in electroanalysis, primarily because of their broad potential window, low background current, rich surface chemistry, low cost, chemical inertness, and suitability for various sensing and detection applications. In contrast, electron transfer rates observed at carbon surfaces are often slower than those observed at metal electrodes. Electron transfer reactivity is strongly affected by the origin and history of the carbon surface (21,22). Numerous studies have thus been devoted for understanding the structure–reactivity relationship at carbon electrodes (21). While all common carbon electrode materials share the basic structure of a six-member aromatic ring and sp^2 bonding, they differ in the relative density of the edge and basal planes at their surfaces. The edge orientation is more reactive than is the graphite basal plane toward electron transfer and adsorption. Materials with different edge-to-basal plane ratios thus display different electron transfer kinetics for a given redox analyte. The edge orientation also displays undesirably high background current contributions. Other factors, besides the surface microstructure, affect the electrochemical reactivity at carbon electrodes. These include the cleanliness of the surface and the presence of surface functional groups. A variety of electrode pretreatment procedures have been proposed to increase the electron transfer rates (21). The type of carbon, as well as the pretreatment method, thus has a profound effect on the analytical performance. The most popular carbon electrode materials are those involving glassy carbon, carbon paste, carbon fiber, screen-printed carbon strips, carbon films, or other carbon composites (e.g., graphite epoxy, wax-impregnated graphite, Kelgraf). The properties of different types of carbon electrodes are discussed below.

4.5.2.2.1 Glassy Carbon Electrodes Glassy (or "vitreous") carbon has been very popular because of its excellent mechanical and electrical properties, wide potential window, chemical inertness (solvent resistance), and relatively reproducible performance. The material is prepared by means of a careful controlled heating program of a premodeled polymeric (phenolformaldehyde) resin body in an inert atmosphere (18). The carbonization process proceeds very slowly over the 300–1200°C temperature range to ensure the elimination of oxygen, nitrogen, and hydrogen. The structure of glassy carbon involves thin, tangled ribbons of cross-linked graphite-like sheets. Because of its high density and small pore size, no impregnating procedure is required. However, surface pretreatment is usually employed to create active and reproducible glassy carbon electrodes and to enhance their analytical performance (19). Such pretreatment is usually achieved by polishing (to a shiny "mirror-like" appearance) with successively smaller alumina particles (down to 0.05 μm). The electrode should then be rinsed with deionized water before use. Additional activation steps, such as electrochemical, chemical, heat, or laser treatments, have also been used to enhance the performance (21). The improved electron transfer capability has been attributed to the removal of surface contaminants, exposure of fresh carbon edges, and an increase in the density of surface oxygen groups (that act as interfacial surface mediators). Several reviews provide more information on the physical and electrochemical properties of glassy carbon electrodes (21,25).

A similar, but yet highly porous, vitreous carbon material, reticulated vitreous carbon (RVC), has found widespread applications for flow analysis and spectroelectrochemistry (26). As shown in Figure 4.10, RVC is an open-pore ("sponge-like") material; such a network combines the electrochemical properties of glassy carbon with many structural and hydrodynamic advantages. These include a very high surface area (~66 cm^2/cm^3 for the 100-ppi grade), 90–97% void volume, and a low resistance to fluid flow.

4.5.2.2.2 Carbon Paste Electrodes Carbon paste electrodes, which use graphite powder mixed with various water-immiscible nonconducting organic binders (pasting liquids), offer an easily renewable and modified surface, low

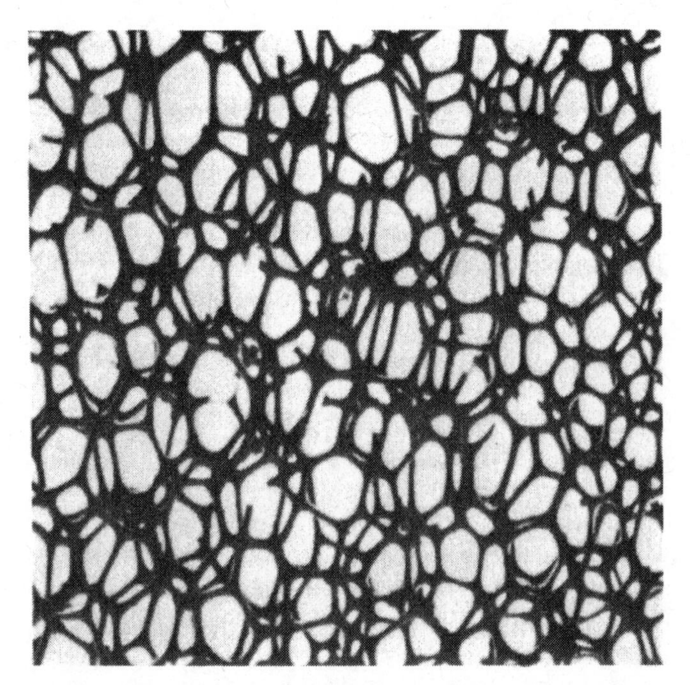

Figure 4.10 The open-pore structure of reticulated vitreous carbon.

cost, and very low background current contributions (27–29). A wide choice of pasting liquids is possible, but practical considerations of low volatility, purity, and economy narrow the choice to a few liquids. These include Nujol (mineral oil), paraffin oil, silicone grease, and bromonaphthalene. The former appears to perform the best. The paste composition strongly affects the electrode reactivity, with the increase in pasting liquid content decreasing the electron transfer rates, as well as the background current contributions (29). In the absence of pasting liquid, the dry graphite electrode yields very rapid electron transfer rates (approaching those of metallic surfaces). Despite their growing popularity, the exact behavior of carbon paste electrodes is not fully understood. It is possible that some of the electrochemistry observed at these electrodes involves permeation of the pasting liquid layer by the electroactive species (i.e., solvent extraction). Carbon paste represents a convenient matrix for the incorporation of appropriate modifying moieties (30). The modifier is simply mixed together with the graphite/binder paste (with no need to devise individualized attachment schemes for each modifier). Enzyme-containing carbon pastes have been used as rapidly responding reagentless biosensors (see Chapter 6). A disadvantage of carbon pastes is the tendency of the organic binder to dissolve in solutions containing an appreciable fraction of organic solvent. Two-dimensional carbon composite electrodes, based on the screen-printing technology, can be prepared from carbon inks consisting of graphite particles, a polymeric binder, and other additives (see Chapter 6).

4.5.2.2.3 Carbon Fiber Electrodes The growing interest in ultramicroelectrodes (Section 4.5.4) has led to a widespread use of carbon fibers in electroanalysis. Such materials are produced, mainly in the preparation of high-strength composites, by high-temperature pyrolysis of polymer textiles or via catalytic chemical vapor deposition. Different carbon fiber microstructures are available, depending on the manufacturing process. They can be classified into three broad categories: low-, medium-, and high-modulus types. The latter is most suitable for electrochemical studies because of its well-ordered graphite-like structure and low porosity (31). Improved electron transfer performance can be achieved by various electrode pretreatments, particularly "mild" and "strong" electrochemical activations, or heat treatment (32). Most electroanalytical applications rely on fibers of 5–20 μm diameter that provide the desired radial diffusion. Such fibers are typically mounted at the tip of a pulled glass capillary with epoxy adhesive, and are used in cylindrical or disk configurations. Precautions should be taken to avoid contamination of the carbon surface with the epoxy. The main advantage of carbon fiber microelectrodes is their small size (5–30 μm diameter of commercially available fibers) that makes them very attractive for anodic measurements in various microenvironments, such as the detection of neurotransmitter release in the extracellular space of the brain. Nanometer-size carbon fibers can be prepared by etching the fiber in flame or under ion beam (e.g., see Fig. 4.11). The various electroanalytical applications of carbon fibers have been reviewed by Edmonds (33).

4.5.2.2.4 Diamond Electrodes Although diamond itself is a known insulator, boron-doped diamond films possess electronic properties ranging from semiconducting to semimetallic and are highly useful for electrochemical measurements. For example, boron doping levels as high as 10^{21} cm^{-3} have been achieved, leading to resistivities lower than 0.01 Ω·cm. Boron-doped diamond (BDD) film electrodes, fabricated by chemical vapor deposition methods, have been studied intensely because of their attractive properties (34,35). These properties include a wide potential window (approaching 3 V, reflecting the high overvoltages for oxygen, chlorine, and hydrogen evolution), low and stable background currents (of a factor of ~10 less than comparably sized polished glassy carbon electrodes), favorable signal: background characteristics, negligible adsorption of organic compounds (i.e., resistance to fouling and a highly stable response), good electrochemical reactivity without any pretreatment, low sensitivity to dissolved oxygen, and extreme hardness. These very low background currents of BDD thin-film electrodes reflect their small double-layer capacitance values, such as 4–8 μF/cm^2 in 1 M KCl over a 2-V window, compared to 25–35 μF/cm^2 under similar conditions at glassy carbon electrodes. This is attributed, in part, to the absence of carbon–oxygen functionalities (35).

Diamond electrodes thus open up new opportunities for work under extreme conditions, including very high anodic potentials, surfactant-rich

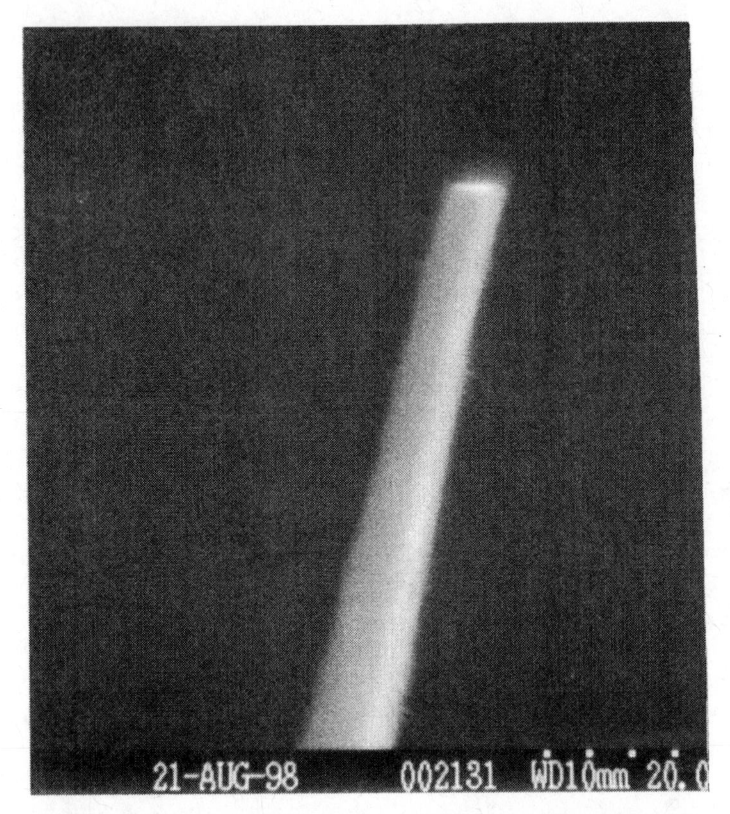

Figure 4.11 Scanning electron image of a carbon fiber electrode.

media, polarization in acidic media, or power ultrasound. Such capabilities and advantages have been illustrated for a wide range of electroanalytical applications, ranging from flow detection of chlorophenols (36) to direct electrochemistry of cytochrome C (37).

4.5.2.3 Metal Electrodes While a wide choice of noble metals is available, platinum and gold are the most widely used metallic electrodes. Such electrodes offer very favorable electron transfer kinetics and a large anodic potential range. In contrast, the low hydrogen overvoltage at these electrode limits the cathodic potential window (to the $-0.2 \rightarrow -0.5$-V region, depending on the pH). More severe are the high background currents associated with the formation of surface oxide or adsorbed hydrogen layers (e.g., see Fig. 4.12). Such films can also strongly alter the kinetics of the electrode reaction, leading to irreproducible data. These difficulties can be addressed with a pulse potential (cleaning–reactivation) cycle, as common in flow amperometry (39). The surface-layer problem is less severe in nonaqueous media, where noble metals

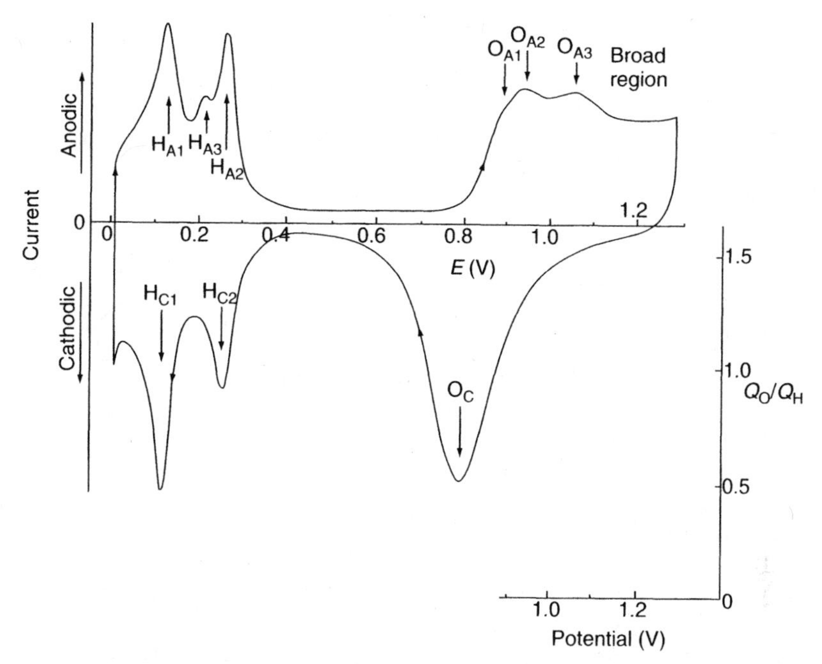

Figure 4.12 Current–potential curve for platinum surface oxide formation and reduction in 0.5 M H_2SO_4. (Reproduced with permission from Ref. 38.)

are often an ideal choice. Compared to platinum electrodes, gold ones are more inert, and hence are less prone to the formation of stable oxide films or surface contamination. Gold electrodes are also widely used as substrates for self-assembled organosulfur monolayers or for stripping measurements of trace metals (Sections 4.5.3 and 3.5; respectively).

Other metals, such as copper, nickel, or silver, have been used as electrode materials in connection with specific applications, such as the detection of amino acids or carbohydrates in alkaline media (copper and nickel) and cyanide or sulfur compounds (silver). Unlike platinum or gold electrodes, these electrodes offer a stable response for carbohydrates at constant potentials, through the formation of high-valence oxyhydroxide species formed in situ on the surface and believed to act as redox mediators (40,41). Bismuth film electrodes (preplated or in situ plated ones) have been shown to be an attractive alternative to mercury films used for stripping voltammetry of trace metals (42,43). Alloy electrodes (e.g., platinum–ruthenium, nickel–titanium) are also being used for addressing adsorption or corrosion effects of one of their components. The bifunctional catalytic mechanism of alloy electrodes (such as Pt–Ru or Pt–Sn ones) has been particularly useful for fuel cell applications (44).

4.5.3 Chemically Modified Electrodes

Chemically modified electrodes (CMEs) represent a modern approach to electrode systems. These electrodes rely on the placement of a reagent onto the surface, to impart the behavior of that reagent to the modified surface. Such deliberate alteration of electrode surfaces can thus meet the needs of many electroanalytical problems, and may form the basis for new analytical applications and different sensing devices. Such surface functionalization of electrodes with molecular reagents has other applications, including energy conversion, electrochemical synthesis, and microelectronic devices.

There are different directions by which CMEs can benefit analytical applications. These include acceleration of electron transfer reactions, preferential accumulation, or selective membrane permeation. Such steps can impart higher selectivity, sensitivity, or stability on electrochemical devices. These analytical applications and improvements have been extensively reviewed (45–47). Many other important applications, including electrochromic display devices, controlled release of drugs, electrosynthesis, fuel cells, and corrosion protection, should also benefit from the rational design of electrode surfaces.

One of the most common approaches for incorporating a modifier onto the surface has been coverage with an appropriate polymer film. Polymer-modified electrodes are often prepared by casting a solution droplet containing the dissolved polymer onto the surface and allowing the solvent to evaporate, by dip or spin coatings, or via electropolymerization in the presence of the dissolved monomer. The latter method offers precise control of the film thickness (and often the morphology), and is particularly attractive in connection with miniaturized sensor surfaces. The structure of some common polymeric coatings is shown in Figure 4.13. For example, the Dupont Nafion perfluorinated sulfonated cation exchanger (Fig. 4.13a) has been widely used as electrode modifier because of its attractive permselective, ion exchange, and antifouling properties (see examples below). In addition to diverse sensing applications, it has been widely used as a proton exchange membrane in fuel cells. Additional advantages can be attained by coupling two (or more) polymers in a mixed or multilayer configuration. Other useful modification schemes include bulk modification of composite carbon materials, covalent (chemical) attachment, sol-gel encapsulation, physical adsorption, and spontaneous chemisorption.

4.5.3.1 Self-Assembled Monolayers

Spontaneously adsorbed monolayers of *n*-alkanethiols [$X(CH_2)_nSH$, with $n > 10$] on gold surfaces, based on the strong interaction between gold and sulfur, are particularly well suited for controlling and manipulating the reactivity at the interface. Such monolayers are commonly formed by immersing the gold electrode in ethanolic solutions containing millimolar concentrations of the alkanethiol overnight. The formation of self-assembled organosulfur monolayers (SAMs) has attracted considerable attention since the late 1980s because of its potential use in many scientific

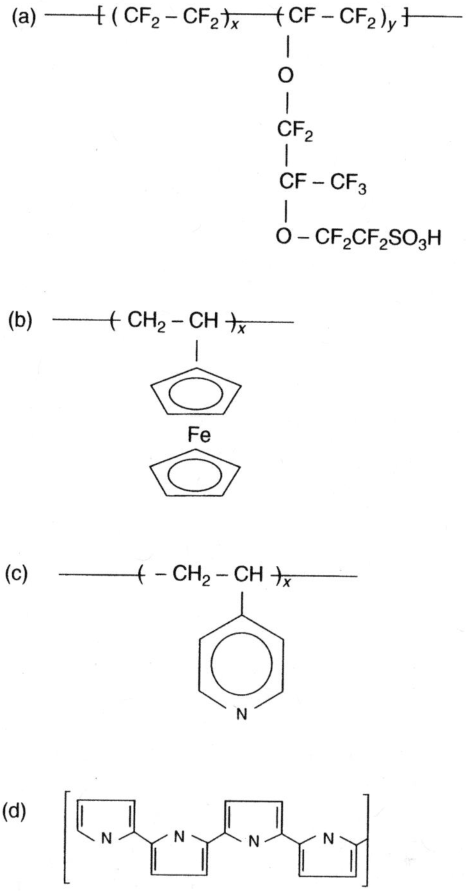

(a) $-[(CF_2-CF_2)_x-(CF-CF_2)_y]-$ with side chain $O-CF_2-CF(CF_3)-O-CF_2CF_2SO_3H$

(b) $-(CH_2-CH)_x-$ Fe

(c) $-(-CH_2-CH)_x-$ N

(d) N N N N

Figure 4.13 Structure of common polymeric coatings: (a) Nafion, (b) polyvinylferrocene; (c) polyvinylpyridine; (d) polypyrrole.

and technological applications (48–50). In addition to fundamental studies on the structure of such monolayers and long-range electron transfer, such applications include chemical sensors and biosensors, information storage devices, or lithography.

Cleavage of the S—H bond is central to this monolayer formation:

$$RSH + Au \rightleftharpoons RS{-}Au + e^- + H^+ \tag{4.9}$$

Van der Waals forces between the methylene groups orient the monolayer. Such a self-assembly process thus results in well-organized and stable monolayers, with the hydrocarbon tails packed parallel to each other, tilted at ~30° relative to the surface normal (Fig. 4.14). The closely-packed pinhole-free films (surface coverage of $\sim 9 \times 10^{-10}\,mol/cm^2$) block transport of species to the

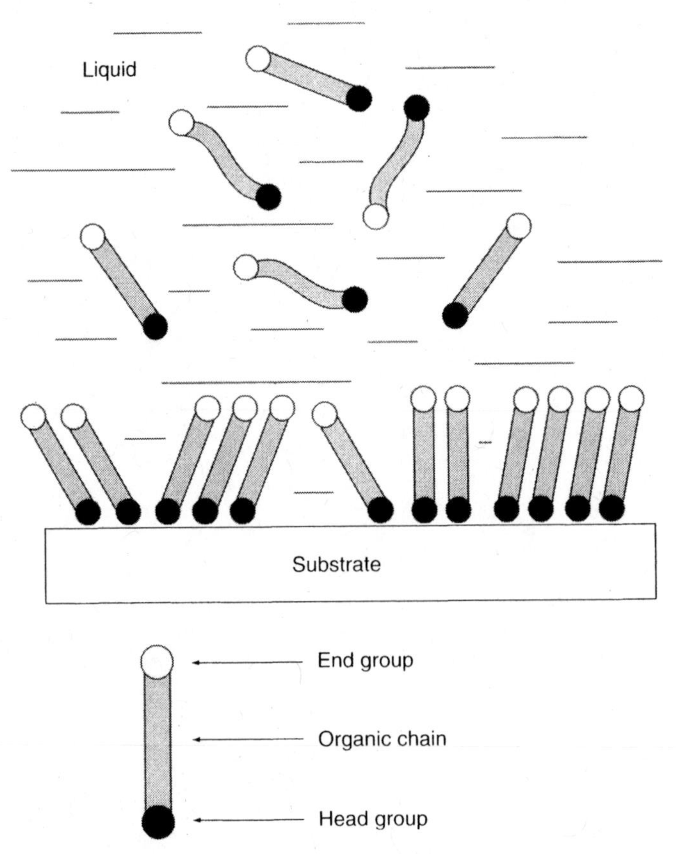

Figure 4.14 Formation of a self-assembled monolayer at a gold substrate. (Reproduced with permission from Ref. 48.)

underlying gold surface. The packing and order are influenced by factors such as chain length, end group, solvent, immersion time, or substrate morphology. Increasingly disordered structures with lower packing density and coverage are observed on decreasing the chain length ($n < 10$). These and other structural disorders and defects (e.g., pinholes) often lead to degraded performance. Coassembled monolayers, formed from mixtures of alkanethiols, can offer compositional and topographical variations in the film architecture. Differences in the coassembled two alkanethiols can be exploited for a selective removal of one component (e.g., by reductive desorption). Patterned SAM nanostructures can be prepared by an AFM-based "dip-pen" lithographic technique (51). Submicrometer SAM patterns can also be prepared by microcontact printing, involving transfer of a pattern from an elastomeric stamp to the gold surface (52). Alkanethiol monolayers can also be assembled on gold nanoparticles confined to carbon electrodes.

The novelty of using SAMs stems from their ability to be further modified into chemically or biologically reactive surface layers (via covalent coupling of different materials to the functional end group, X). Such use of SAM for anchoring various functionalities can impart specific interactions essential for various sensing applications (see Chapter 6). SAMs of DNA oligonucleotides on gold electrodes have been particularly useful for the development of DNA hybridization biosensors (see Section 6.1.2.2). These applications can benefit from the use of mixed SAMs (e.g., monolayers of thiol-derivatized single-stranded oligonucleotide and 6-mercapto-1-hexanol) that minimize steric/hindrance effects. The high degree of order of SAMs has also allowed the development of detailed structure–function relationships for various electron transfer processes. The well-defined surfaces of SAM-modified electrodes have also been extremely useful for studying electron transfer rates of proteins (53). As expected, the rate constants have been found to depend on the length of the spacer (i.e., the donor–acceptor distance). Other highly ordered films, such as those of alkyl siloxane, can be formed on metal oxide surfaces (particularly SiO_2).

4.5.3.2 Carbon-Nanotube-Modified Electrodes Carbon nanotubes (CNTs) represent an increasingly important group of nanomaterials with unique geometric, mechanical, electronic, and chemical properties (54). CNTs can be divided into single-wall carbon-nanotubes (SWCNT) and multiwall carbon nanotubes (MWCNTs). SWCNTs possess a cylindrical nanostructure (with a high aspect ratio), formed by rolling up a single graphite sheet into a tube. Multiwall CNTs consist of an array of such nanotubes that are concentrically nested like rings of a tree trunk. The unique properties of CNT make them also extremely attractive for the tasks of surface modification and electrochemical detection. More recent studies have demonstrated that CNT-modified electrodes can promote the electrochemical reactivity of important analytes and impart resistance against surface fouling (55). The electrocatalytic activity of CNT has been attributed to the presence of edge plane defects at their end caps (56). "Trees" of aligned CNT in the nanoforest, prepared by self-assembly (Fig. 4.15), can act as molecular wires to allow electrical communication between the underlying electrode and redox proteins (covalently attached to the ends of the SWNT) (57,58). Such efficient electron transfer to enzyme redox centers offers great promise for the design of amperometric biosensors (59).

4.5.3.3 Sol-gel Encapsulation of Reactive Species Another new and attractive route for tailoring electrode surfaces involves the low-temperature encapsulation of recognition species within sol-gel films (60,61). Such ceramic films are prepared by the hydrolysis of an alkoxide precursors, such as $Si(OCH_3)_4$, under acidic or basic condensation, followed by polycondensation of the hydroxylated monomer to form a three-dimensional interconnceted porous network. The resulting porous glass-like material can physically retain

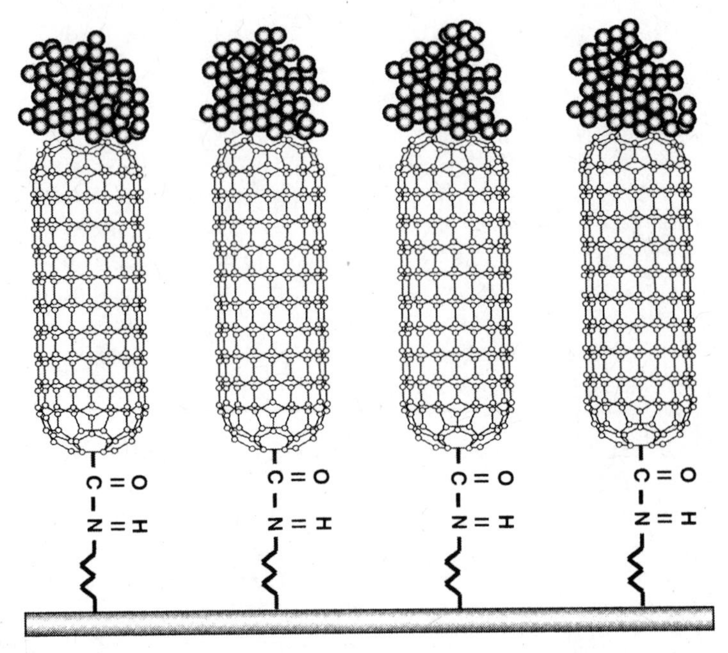

Figure 4.15 Nanoforest of vertically aligned CNT "trees" acting as molecular wires. (Reproduced with permission from Ref. 57.)

the desired modifier, but permits its interaction with the analyte that diffuses into the matrix. Besides their ability to entrap the modifier, sol-gel processes offer tunability of the physical characteristics (e.g., porosity), thermal stability, and mechanical rigidity. Sol-gel-derived composite electrodes have also been prepared by dispersing carbon or gold powders in the initial sol-gel mixture (62,63). Highly selective electrochemical sensors can be developed by application of molecular imprinting in sol-gel films, involving the formation of selective binding for the target analyte (64).

4.5.3.4 Electrocatalytically Modified Electrodes Often the desired redox reaction at the bare electrode involves slow electron transfer kinetics and therefore occurs at an appreciable rate only at potentials substantially higher than its thermodynamic redox potential. Such reactions can be catalyzed by attaching to the surface a suitable electron transfer mediator (65,66). Electrocatalytic reactions play a central role in electrochemistry and a vital role in sensing and energy-related applications. Knowledge of homogeneous solution kinetics is often used for selecting the surface-bound catalyst. The function of the mediator is to facilitate the charge transfer between the analyte and the electrode. In most cases the mediated reaction sequence (e.g., for a reduction process) can be described by

$$M_{ox} + ne^- \rightarrow M_{red} \tag{4.10}$$

$$M_{red} + A_{ox} \rightarrow M_{ox} + A_{red} \tag{4.11}$$

where M represents the mediator and A, the analyte. Hence, the electron transfer takes place between the electrode and mediator and not directly between the electrode and the analyte. The active form of the catalyst is electrochemically regenerated. The net results of this electron shuttling are a lowering of the overvoltage to the formal potential of the mediator and an increase in current density. The efficiency of the electrocatalytic process also depends on the actual distance between the bound redox site and the surface (since the electron transfer rate decreases exponentially when the electron–tunneling distance is increased).

The improvements in sensitivity and selectivity accrue from electrocatalytic CMEs have been illustrated for numerous analytical problems, including the biosensing of dihydronicotinamide adenine dinucleotide (NADH) at a Meldola Blue–coated electrode (67), the liquid chromatographic amperometric detection of thiols at cobalt phthalcocyanine–coated electrodes (68), detection of nitric oxide release from a single cell by an electropolymerized nickel(II)-porphyrinic-based carbon fiber microsensor (69), flow injection measurements of carbohydrates at ruthenium dioxide–containing carbon paste detectors (70), and low-potential detection of hydrogen peroxide at Prussian Blue–modified electrodes (71). Cyclic voltammograms for various carbohydrates at the ruthenium dioxide carbon paste electrodes are shown in Figure 4.16. As expected for redox mediation, the peaks of the surface-bound ruthenium species (dashed lines) increase on addition of the carbohydrate analytes (solid lines). Figure 4.17 illustrates the electrocatalytic scheme involved in the detection of NADH. The implications of this scheme on various biosensors are discussed in Section 6.1. Electrocatalytic surfaces are also widely used in energy-producing fuel cells, particularly for catalyzing the oxidation of methanol or the reduction of oxygen (72).

4.5.3.5 Preconcentrating Electrodes Preconcentrating CMEs, with surfaces designed for reacting and binding of target analytes, hold great promise for chemical sensing (73–76). The concept is analogous to stripping voltammetric schemes, where the target analyte is preferentially partitioned from the dilute sample into the preconcentrating surface layer, and is subsequently reduced or oxidized during a potential scan. Unlike conventional stripping procedures, the preconcentration step is nonelectrolytic. Most preconcentrating CMEs employ electrostatic binding or coordination reactions, for collecting the analyte. Schemes based on hydrophobic partition into a lipid coating, covalent reactions, or peptide binding have also been reported. The preconcentrating agent may be incorporated within the interior of a carbon paste matrix or via functionalized polymeric and alkanethiol films. For example, as shown in Figure 4.18, ligand centers can covalently bind to a polymer back-

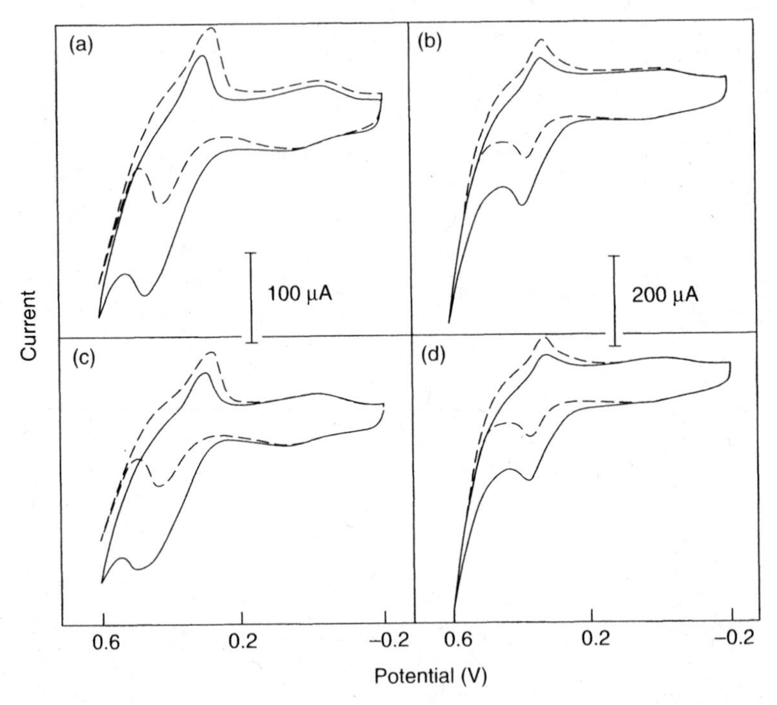

Figure 4.16 Cyclic voltammograms for 1.5×10^{-3} M ribose (a), glucose (b), galactose (c), and fructose (d) recorded at a RuO_2-modified carbon paste electrode. Dotted lines were obtained in carbohydrate-free solutions. (Reproduced with permission from Ref. 70.)

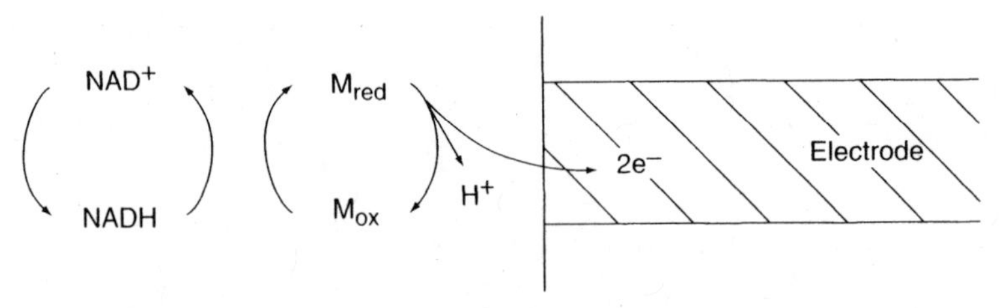

Figure 4.17 Electrocatalytic detection of NADH.

bone on the electrode to effectively accumulate and measure target metals. The major requirements for a successful analytical use of preconcentrating electrodes are strong and selective binding, prevention of saturation, and a convenient surface regeneration. Following the accumulation, the electrode can be transferred to more suitable solutions that facilitate the measurement and "cleaning" steps.

Practical examples of using preconcentrating CMEs include the use of a mixed 2,9-dimethyl-1,10-phenanthroline/carbon paste electrode for trace

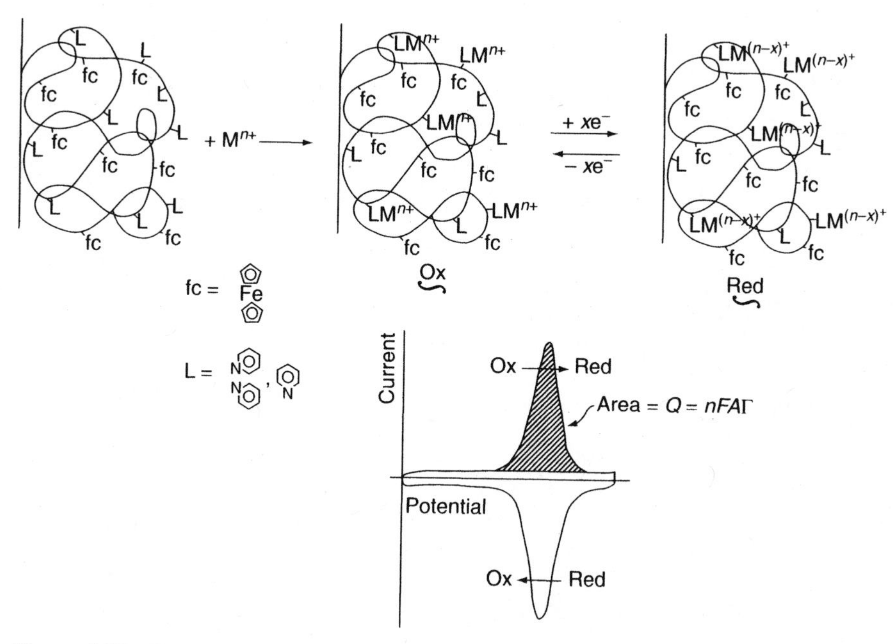

Figure 4.18 Preconcentrating surfaces based on covalent binding of the ligand to a polymer backbone. (Reproduced with permission from Ref. 74.)

measurements of copper (77), the use of clay-containing carbon pastes for voltammetric measurements of iron (78), the use of polyelectrolyte coatings for the uptake and voltammetry of multiple-charged metal complexes (79,80), the use of surface-bound crown ethers and cryptands for trace measurements of lead (81), the ion exchange voltammetric measurements of lanthanide ions at a Nafion-coated electrode (82), the collection of ultratrace cadmium onto mercaptocarboxylic acid monolayers (83), ultrasensitive biomimetric detection of copper based on a selective surface-confined peptide ligand (84), and the quantitation of nickel at porphyrin-coated electrodes (85).

4.5.3.6 Permselective Coatings Permselective coatings offer the promise of bringing higher selectivity and stability to electrochemical devices. This is accomplished by exclusion from the surface of unwanted matrix constituents, while allowing transport of the target analyte. Different avenues to control the access to the surface, based on different transport mechanisms, have been proposed. These include the use of size-exclusion poly(1,2-diaminobenzene) films (86), charged-exclusion ionomeric Nafion coatings (87), hydrophobic lipid (88) or alkanethiol (89) layers, or bifunctional (mixed) coatings (90). Such anti-interference membrane barriers offer an effective separation step (in situ on the surface), and hence protect the surface against adsorption of large macromolecules or minimize overlapping signals from undesired electroactive interferences. For example, the poly(1,2-diaminobenzene)-coated flow detector rapidly responds to the small hydrogen peroxide molecule, but not to the larger ascorbic acid, uric acid, or cysteine species (Fig. 4.19). Note also the protection from foulants present in the serum sample. Such size exclusion (sieving) properties are attributed to the morphology of electropolymerized films (Fig. 4.20). Similarly, Nafion-coated microelectrodes are often used for in

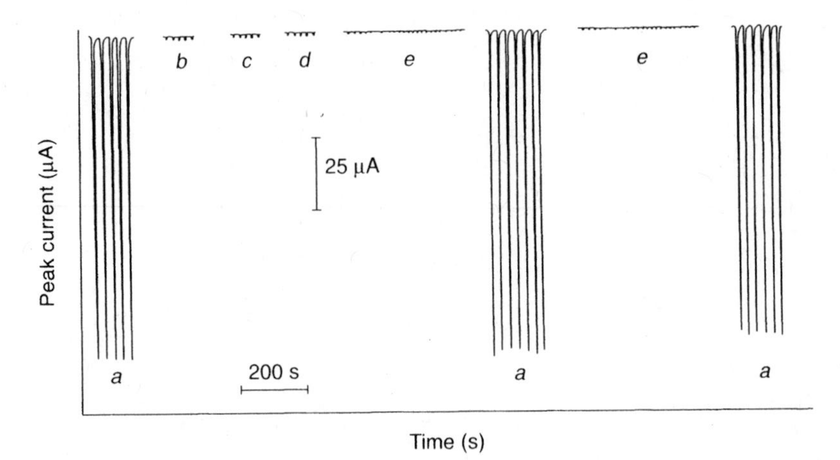

Figure 4.19 Permselective coatings: flow injection response of a poly (1,2-diaminobenzene)-coated electrode to the following: *a*, hydrogen peroxide (1 mM); *b*, ascorbic acid (1 mM); *c*, uric acid (1 mM); *d*, L-cysteine (1 mM); *e*, control human serum. (Reproduced with permission from Ref. 86.)

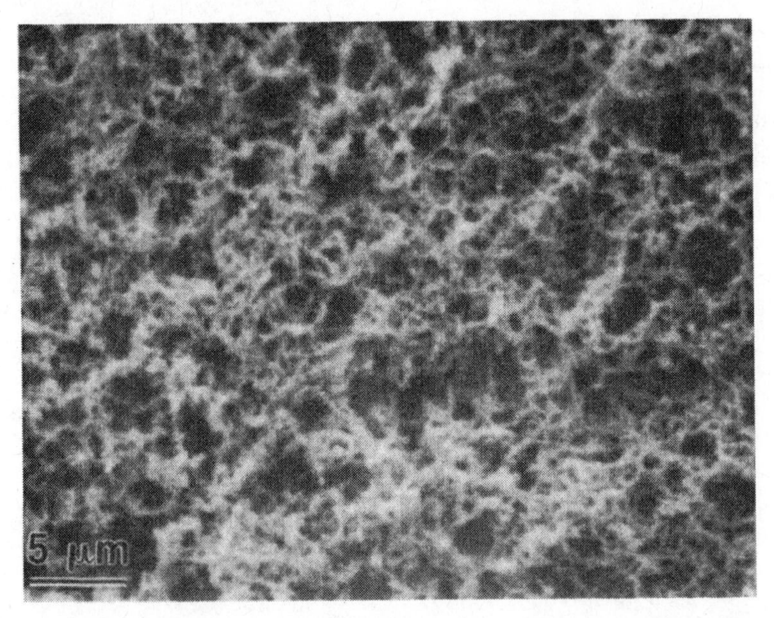

Figure 4.20 Scanning electron micrograph of a polyaniline-coated electrode.

vivo monitoring of cationic neurotransmitters, such as dopamine, in the brain extracellular fluid in the presence of otherwise interfering ascorbic acid (87). Such anionic interference is excluded from the surface through electrostatic repulsion with the negatively charged sulfonated groups (Fig. 4.21). Examples of these and other discriminative films are given in Table 4.2.

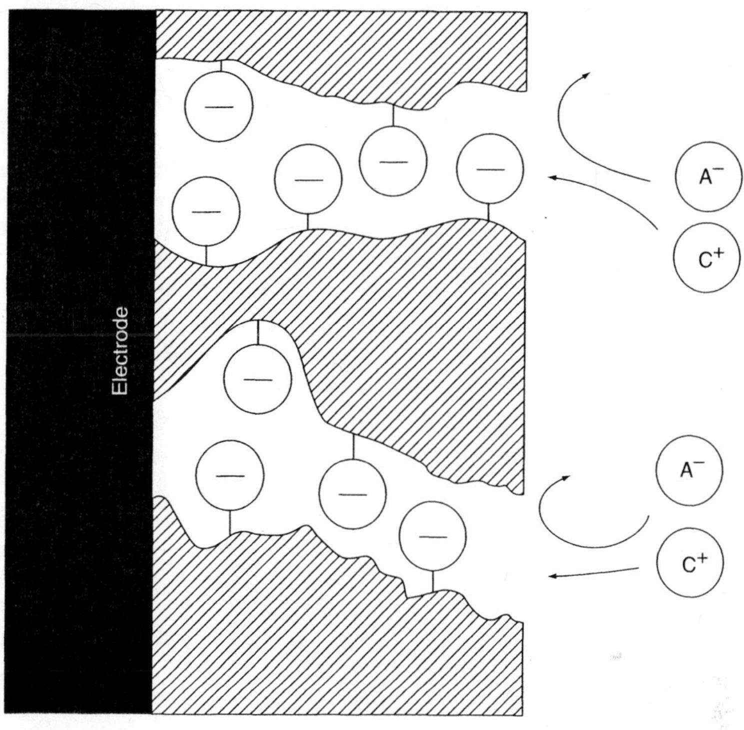

Figure 4.21 Use of negatively charged polymeric films for excluding anionic interferences.

TABLE 4.2 Commonly Used Membrane Barriers

Transport Mechanism	Membrane Barrier	Ref.
Size exclusion	Cellulose acetate	91
	Poly(1,2-diaminobenzene)	86
	Polyphenol	92
Hydrophobic barriers	Phospholipid	88
	Self-assembled thiols	89
Charge exclusion	Nafion	87
	Poly(ester sulfonic acid)	93
	Self-assembled thioctic acid	94
Mixed control	Cellulose acetate/Nafion	90

4.5.3.7 Conducting Polymers Electronically conducting polymers (such as polypyrrole, polythiophene, and polyaniline) have attracted considerable attention because of their ability to reversibly switch between the positively charged conductive state and a neutral, essentially insulating form, and to incorporate and expel anionic species (from and to the surrounding solution), on oxidation or reduction:

$$P^0 + A^- \rightarrow P^+A^- + e^- \tag{4.12}$$

where P and A^- represent the polymer and the "dopant" anion, respectively. The latter serves to maintain the electrical neutrality, counterbalancing the positive charge of the polymer backbone. The redox changes [Eq. (4.12)] are not localized at a specific center, but rather delocalized over a number of conducting polymer groups. The conjugation of the π-electron system creates a molecular orbital which extends throughout the polymer chain. The electrical conductivity of these films, which originates from the electronic structure of their polymeric backbone (i.e., electron hopping involving the delocalized π electrons) can vary with the applied potential. The conductivity values depend on the amount of carriers (electrons or holes) created in the polymer chain (which is determined by the concentration of dopant in the polymer) and the carrier mobility through the polymer. The structure of common conducting polymers, and their conductivity range (from the undoped to doped states) are displayed in Figure 4.22. Such conductivity values are similar to those of inorganic semiconductors. It is possible also to use large organic anions (and even DNA) as dopants. Such large dopants cannot be readily expelled from the polymer network. As a result, when the polymeric chain becomes neutral (in the reduced state) the negative charge of the entrapped anionic dopant is balanced by insertion of the electrolyte cation.

These polymers are readily prepared by in situ electropolymerization (from the monomer solution). Oxidation of the monomer proceeds according to

$$\text{(monomer ring, X)} \xrightarrow{E_{app}} \left[\text{(chain of four rings, X)} \right]^{+} A^- \tag{4.13}$$

X = NH (polypyrrole)
X = S (polythiophene)
X = O (polyfuran)

Often the first step in the electropolymerization process is the electrooxidative formation of a radical cation from the starting monomer. This step is commonly followed by a dimerization process, followed by further oxidation and coupling reactions. Well-adhered films can thus be formed on the surface in galvanostatic, potentiostatic, or multiscan experiments. The behavior of elec-

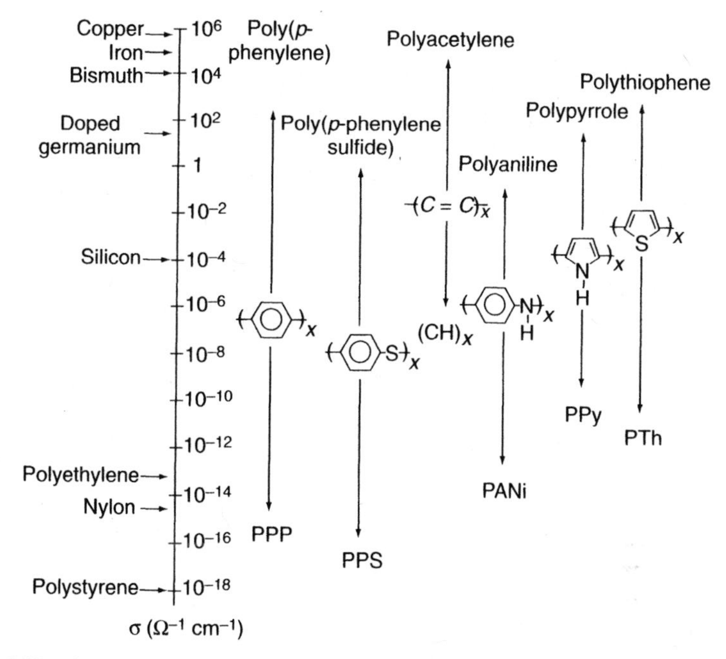

Figure 4.22 Conductivity ranges of common conducting polymers, along with their chemical structures. (Reproduced with permission from Ref. 95.)

tropolymerized films can be controlled by the polymerization conditions, including the electrolyte (particularly the nature and level of the anion serving as the dopant), solvent, monomer concentration, applied potential or current, and duration. The dynamic of the redox switching reaction [Eq. (4.12)] strongly depends on the ionic fluxes that accompany the process. As was discussed earlier, the tight entrapment of large anionic dopants (e.g., polyelectrolytes) precludes their removal, and hence the charge compensation is dominated by the movement of a "pseudodopant" cation.

Changes in the polymer properties can be induced by attaching various chemical or biological functionalities to the monomer prior to polymerization. A large variety of functions can thus be incorporated into conducting polymers owing to the richness of chemical synthesis. It is also possible to impart molecular recognition or electrocatalytic action via the incorporation of functional dopants (e.g., complexing agents or an electron transfer mediator). Hence, conducting polymers can act as efficient molecular interfaces between recognition elements and electrode transducers. The unique physical and chemical properties of conducting polymers, particularly the controllable and dramatic change in electrical conductivity and rapid exchange of the doping ion, offer diverse electrochemical applications, including batteries, fuel cells, corrosion protection, or chemical sensing (96). The latter include amperometric flow detection of nonelectroactive ions, solid-state gas sensing, entrapment/attachment and stabilization of biological entities, direct monitoring of biological interactions (e.g., DNA hybridization and antibody–antigen binding), electrochemical control of membrane permeability, sensor arrays based on multiple films, new potentiometric recognition of ions, preconcentration/stripping of trace metals, and controlled release of chemicals. For example, Figure 4.23 shows typical flow injection peaks for submillimolar concentrations of carbonate ions (utilizing the doping-updoping of the polypyrrole-based detector). Such anodic peaks reflect the passage of the carbonate anions [which are capable of penetrating into the film, equation (4-12)] over the surface. These and other analytical opportunities have been reviewed by Ivaska (98), Teasdale and Wallace (99), and Bidan (95). The widespread use of polypyrrole is attributed to its electropolymerization from aqueous media at neutral pH (which allows the entrapment of a wide range of dopants). Other films are more limited in this regard. For example, thiophene is soluble only in organic solvents and aniline, only in acidic media. The electropolymerization growth can also lead to nonconducting, self-limiting films, which are often used as permselective/protective layers (Section 4.5.3.5) or for the physical entrapment of biomolecules (Chapter 6).

More recently introduced conducting polymer nanowires are characterized by higher surface : volume ratio (i.e., larger aspect ratio) compared to conventional conducting polymer films, and hence offer greater promise for resistive sensors and molecular electronic devices (100). Such scaling down of conducting polymer films into nanometer-scale wires leads to highly sensitive fast-responding sensors. Conducting polymer nanowires of controlled dimension can be readily prepared by a template-directed electrochemical synthesis involving electrodeposition into the pores of a membrane template or within the microchannels contacting neighboring microelectrodes (101,102).

Electropolymerization can also be used for the design of molecularly imprinted polymers (MIPs), capable of interacting with the analyte (template) molecule with high affinity and specificity (103,104). This is accomplished by electropolymerizing polypyrrole, polyaniline, or poly(*o*-phenylenediamine) in the presence of the analyte (template) molecule. At the end of the polymer-

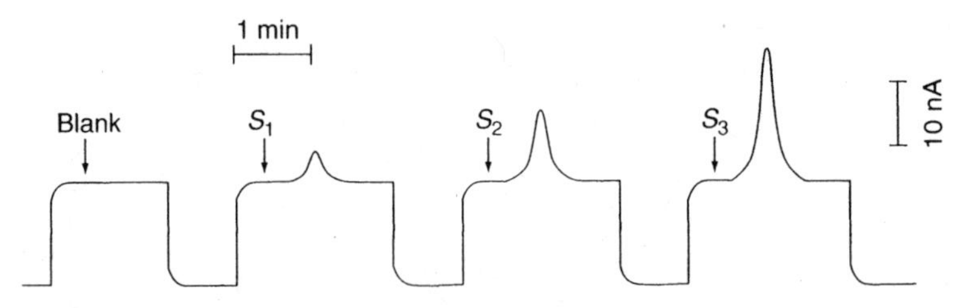

Figure 4.23 Typical response of polypyrrole detector to carbonate: S_1, 1×10^{-4} M; S_2, 2.5×10^{-4} M; S_3, 5×10^{-4} M, based on the doping–undoping process. (Reproduced with permission from Ref. 97.)

ization, the template is removed leaving a crosslinked polymer, containing sites with high affinity for the template (Fig. 4.24). Such electrical preparation of MIP is extremely attractive for the design of sensors with high molecular recognition capability. More details on MIP-based electrochemical sensors are given in Chapter 6.

4.5.4 Microelectrodes

Miniaturization is a growing trend in the field of analytical chemistry. The miniaturization of working electrodes not only has obvious practical advantages but also creates some fundamentally new possibilities (105–108). The term "microelectrodes" is reserved here for electrodes with at least one dimension not greater than 25 μm.

Such dimensions offer obvious analytical advantages, including the exploration of microscopic domains, measurements of local concentration profiles, detection in microflow systems, and analysis of very small (microliter) sample volumes. Particularly fascinating are more recent studies aimed at time-resolved probing of dynamic processes (e.g., secretion of chemical messengers) in single cells (109), the in vivo monitoring of neurochemical events (e.g., stimulated dopamine release) (110), the in vivo detection of nitric oxide (111), use of nanoscopic electrode tips for single-molecule detection (112), or high-resolution spatial characterization of surfaces (see Section 2-3). Figure 4.25 illustrates the use of a carbon fiber microelectrode for measuring the vesicular release of dopamine following cellular stimulation. Microelectrodes with 1 μm diameters provide spatial resolution sufficient to identify the locations of release sites on the surface of single cells (110).

Microelectrodes exhibit several other attractive and important properties that have expanded the possibilities of electrochemistry:

1. Because of the very small total currents at microelectrodes, it is possible to work in highly resistive solutions that would develop large ohmic (iR) drops

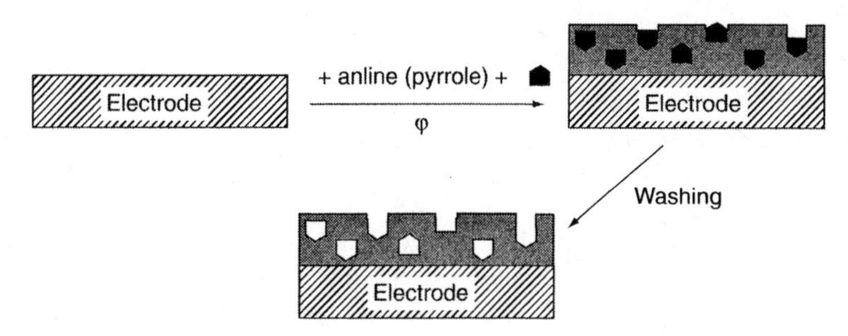

Figure 4.24 Use of electropolymerization from preparing molecularly imprinted polymers with sites with high affinity for the template (analyte) molecule. (Reproduced with permission from Ref. 103.)

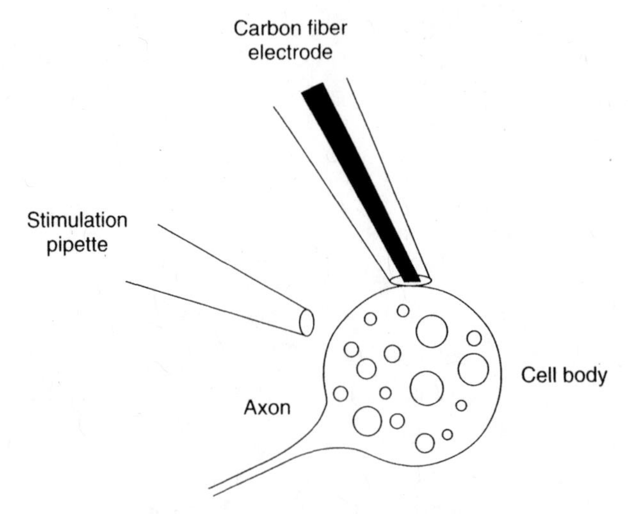

Figure 4.25 Experimental setup for monitoring dopamine release by exocytosis, from a cell body. The microelectrode and glass capillary (containing the chemical stimulant) are micromanipulated up to the cell body. (Reproduced with permission from Ref. 113.)

with conventional electrodes. The decreased ohmic distortions allow electrochemical measurements to be made on new and unique chemical environments (which are not amenable to macroscopic electrodes). Microelectrode experiments have thus been reported in low dielectric solvents (e.g., benzene, toluene), frozen acetonitrile, low-temperature glasses, gaseous and solid phases, supercritical carbon dioxide, ionically conductive polymers, oil-based lubricants, and milk. In addition, more traditional systems can be studied with little or no deliberately added supporting electrolyte, and with two-electrode systems. The use of electrolyte-free organic media can greatly extend the electrochemical potential window, thus allowing studies of species with very high oxidation potentials. Acetonitrile, for example, can be used to about 4 V (vs. a silver reference electrode), making possible studies of short-chain alkanes. Work without deliberately added supporting electrolyte is advantageous also for stripping voltammetry of trace metals as it minimizes potential impurities and changes in the chemical equilibrium.

2. The greatly reduced double-layer capacitance of microelectrodes, associated with their small area, results in electrochemical cells with small *RC* time constants. For example, for a microdisk the *RC* time constant is proportional to the radius of the electrode. The small *RC* constants allow high-speed voltammetric experiments to be performed at the microsecond timescale (scan rates higher than 10^6 V/s) and hence to probe the kinetics of very fast electron transfer and coupling chemical reactions (114) or the dynamic of processes such as exocytosis (e.g., Fig. 4.25). Such high-speed experiments are discussed further in Section 2.1.

3. Enhanced rates of mass transport of electroactive species accrue from the radial (nonplanar) diffusion to the edges of microelectrodes. Such "edge effects" contribute significantly to the overall diffusion current. The rate of mass transport to and from the electrode (and hence the current density) increases as the electrode size decreases. As a consequence of the increase in mass transport rates and the reduced charging currents, microelectrodes exhibit excellent signal-to-background characteristics in comparison to their large counterparts. In addition, steady-state or quasi-steady-state currents are rapidly attained, and the contribution of convective transport is negligible. The fact that redox reactions that are limited by mass transport at macroscopic electrodes become limited by the rate of electron transfer can also benefit many kinetic studies (115).

4.5.4.1 Diffusion at Microelectrodes The total diffusion-limited current is composed of the planar flux and radial flux diffusion components:

$$i_{\text{total}} = i_{\text{planar}} + i_{\text{radial}} \tag{4.14}$$

For disk, spherical, and hemispherical geometries, the general expression for the radial component in Eq. (4.14) is given by

$$i_{\text{radial}} = arnFDC \tag{4.15}$$

where r is the electrode radius and a is a function of the electrode geometry. For disks, spheres, and hemispheres the a values are equal to 4, 4π, and 2π, respectively. Such radial diffusion leads to a larger flux at the perimeter of the electrode (than at the center), and hence to a nonuniform current density.

The extent by which the planar or radial component dominates depends on the relative dimensions of the electrode and the diffusion layer, as expressed by the dimensionless parameter Dt/r_0^2, where t is the electrolysis time, and r_0 is the smallest dimension of the electrode (116). For large (>1) values of Dt/r_0^2 (i.e., diffusion-layer thickness that exceeds the size of the electrode), the current approaches steady state, and sigmoidal voltammograms are observed. In contrast, planar diffusion dominates at small values of Dt/r_0^2, and a peak-shaped behavior is observed. Hence, depending on the timescale of the experiment (i.e., the scan rate), the same electrode may exhibit peak-shaped or sigmoidal voltammograms (e.g., see Fig. 4.26). Similarly, in the case of chronoamperometric experiments, a modified Cottrell equation predicts that a steady-state current is reached rapidly after the potential step (e.g., within ~10 ms and 1.3 s for 1- and 10-μm-diameter disks, respectively). The change from semiinfinite planar diffusion into semiinfinite hemispherical diffusion, associated with the decrease in the electrode dimension, is illustrated in Figure 4.27, which displays computed concentration profiles for a given time after the start of a chronoamperometric experiment at a disk with different radii.

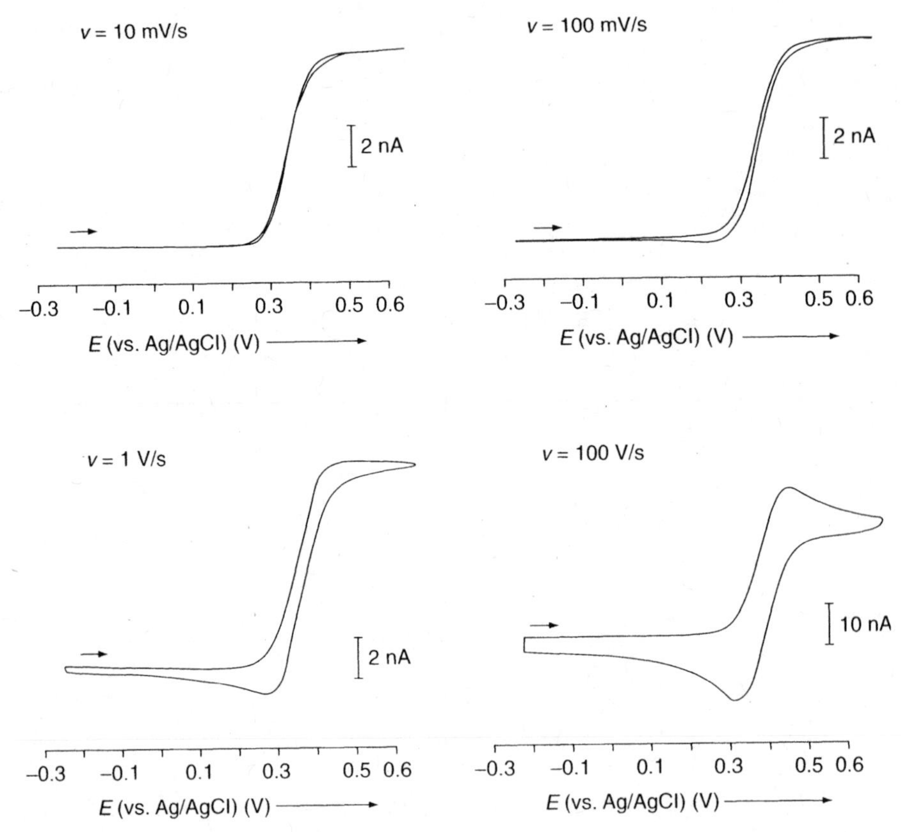

Figure 4.26 Cyclic voltammograms for the oxidation of ferrocene at a 6-μm platinum microdisk at different scan rates. (Reproduced with permission from Ref. 116.)

4.5.4.2 Microelectrode Configurations Electrodes of different materials have been miniaturized in many geometric shapes (Fig. 4.28), with the common characteristic that the electrode dimension is significantly smaller than the diffusion layer at the electrode surface (for ordinary voltammetric time scales, e.g., 1–10 s). The most commonly used shape is a circular conductor (of around 10 μm diameter), embedded in an insulating plane (the so-called microdisk electrode) (Fig. 4.28a). Other common geometries include the microring (Fig. 4.28b), microcylinder (Fig. 4.28c), microhemisphere (Fig. 4.28d), or microband (Fig. 4.28e) electrodes. Cylinder and band (line) microelectrodes, which can be several millimeters long, yield larger (and hence more easily measured) currents, while maintaining an enhanced diffusional flux. Band electrodes of nanoscopic dimensions can be fabricated by sealing ("sandwiching") ultrathin carbon and metal films between insulating supports and polishing one end of the sandwich or via photolithographic deposition of a thin metal film on an insulating substrate. The fabrication of most

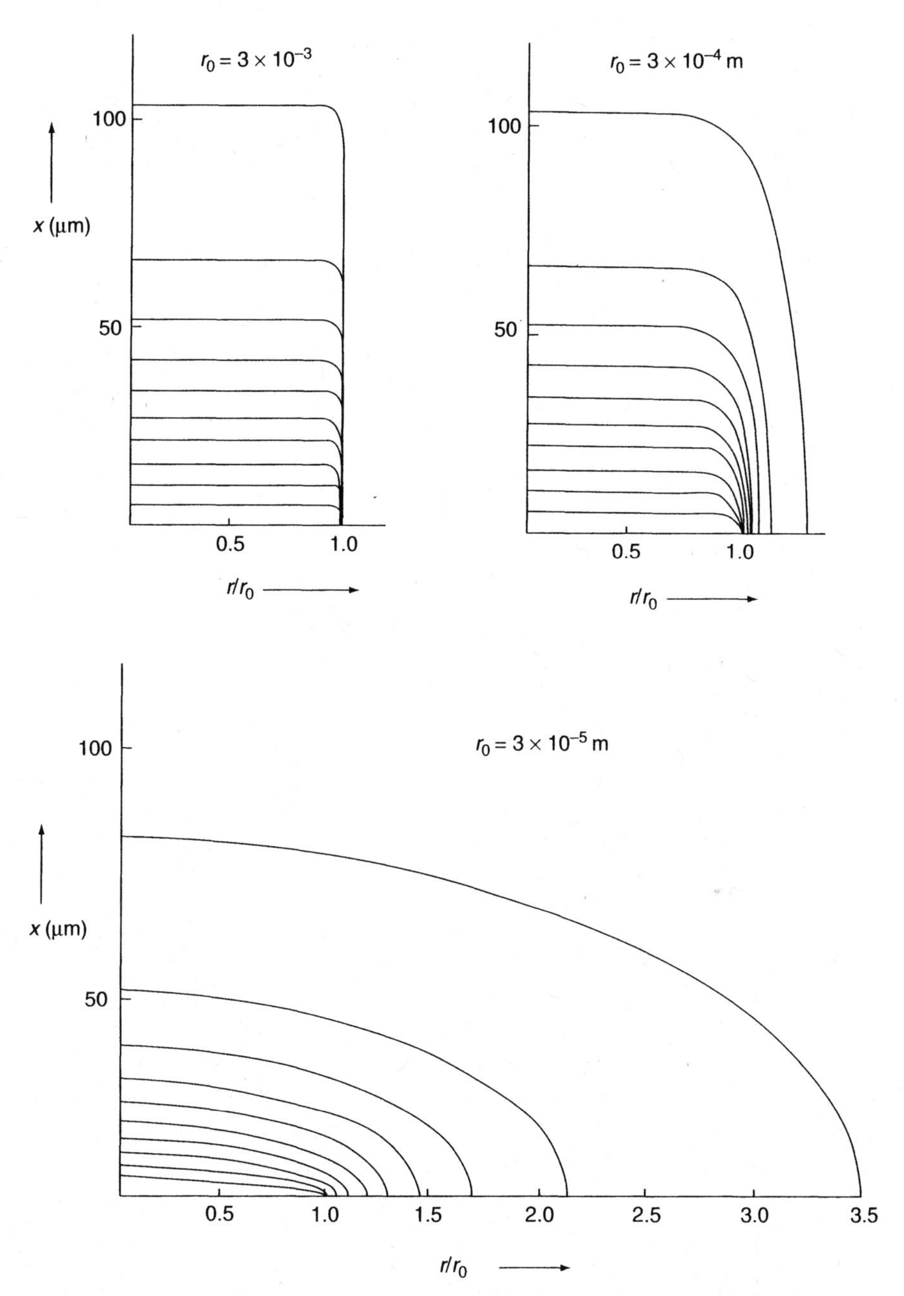

Figure 4.27 Normalized calculated concentration profiles for disk electrodes with different radii (r_0), 1 s after start of a chronoamperometric experiment. (Reproduced with permission from Ref. 116.)

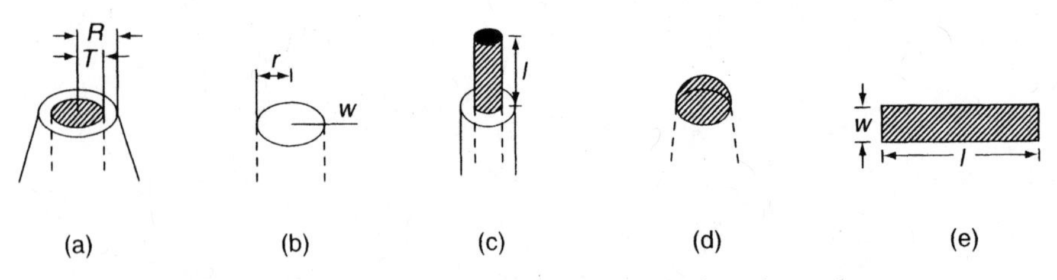

Figure 4.28 Common configurations of microelectrodes.

microelectrode geometries (with the exception of microcylinders) is technically demanding (108). Special attention should be given to proper sealing (between the active surface and insulating sheath) to ensure good performance and to minimize stray capacitances. Fine metal (Pt, Au, Ir) wires, carbon fibers, or thin metal films are commonly used for these preparations. Once fabricated, such ultramicroelectrodes are generally characterized by scanning electron microscopy, steady-state voltammetry, or scanning electrochemical microscopy (108). Molecule (nanometer)-sized electrodes, which have been developed in several laboratories (117–120), should offer additional spatial and temporal advantages, including a closer look at interfacial chemistry processes and measurements of fast electron transfer rates not accessible using larger microelectrodes. This is attributed to the very rapid rates of mass transport at nanoscale electrodes. The construction of such nanoscopic electrodes is even more challenging than their microscopic counterparts (121). Nanometer-sized (2–150-nm) platinum electrodes, fabricated by electrophoretic coating of etched platinum wire with poly(acrylic acid), have been shown useful for detecting as few as 7000 molecules (118).

4.5.4.3 Composite Electrodes Composite electrodes couple the advantages of single-microelectrode systems with significantly higher currents due to larger surface areas (122). Such electrodes thus address instrumental difficulties of monitoring the extremely small (subnanoampere) currents at single microelectrodes. The surface of composite electrodes consists of uniform (array) or random (ensemble) dispersion of a conductor region within a continuous insulating matrix (Fig. 4.29). Examples of arrays include closely spaced microdisks or interdigitated microband electrodes (e.g., Fig. 4.30). Lithographic techniques are often used for fabricating such arrays (with different patterning) allowing precise control of the spacing (123,124). Ensembles can be fabricated by mixing/pressing a powdered conductor with an insulator [e.g., Kel-F/graphite (125)], by impregnation of a porous conductor with an insulator [e.g., microcellular carbon/epoxy (126)], embedding carbon fibers in an insulating epoxy (127), or via deposition of a metal conductor into the pores of a microporous host membrane (101,128). The latter also represents an attractive route for preparing nanowires for a wide range of nanotechnology applications (in connection to dissolution of the membrane template) (129).

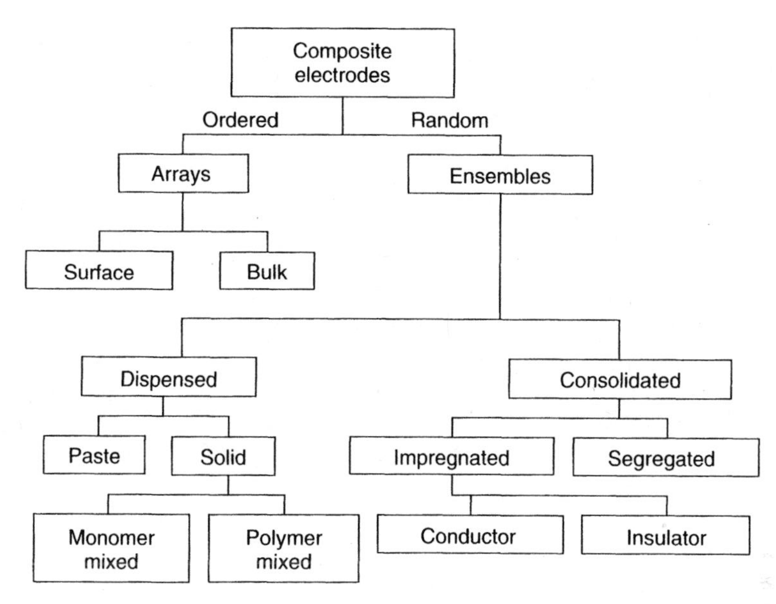

Figure 4.29 Classification of composite electrodes used in controlled-potential electrochemical techniques. (Reproduced with permission from Ref. 122.)

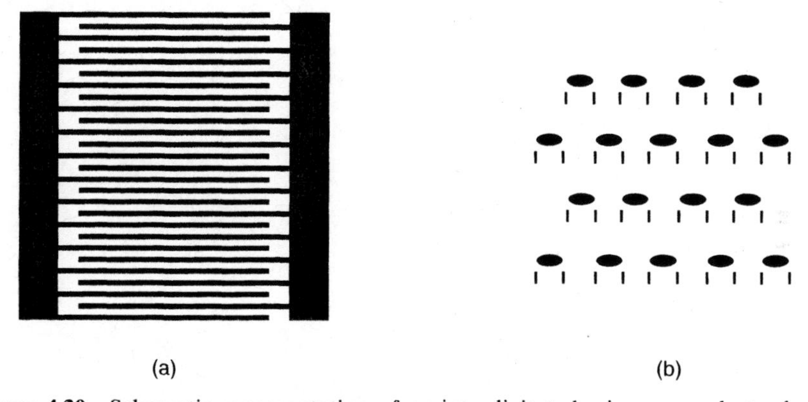

Figure 4.30 Schematic representation of an interdigitated microarray electrode (a) and closely spaced microdisk electrodes (b).

As long as there is a negligible overlap of the diffusion layers from adjacent sites (i.e., each member maintains its own radial diffusional field), the current of composite electrodes is the sum of currents of the individual sites. At sufficiently long times, the diffusion layers overlap, and the electrode behaves as though its entire geometric surface were active. For example, a slow scanning cyclic voltammetric experiment displays a current peak proportional to the total geometric area. The exact timescale for the change from isolated to merged diffusion layers depends on the spacing between the individual electrodes and their size. Larger distances and smaller dimensions are preferred. Chronoamperometric experiments, such as the one shown in Figure 4.31, can be used to estimate the transition between these time regimes, and the fraction of the conductive surface in accordance to the theoretical model (130). In addition to their large collective current, enhanced signal-to-noise ratios, and flow rate independence (in flow detection), composite electrodes hold great promise for incorporating appropriate modifiers within the bulk of the composite matrix.

Closely spaced band electrodes (pairs or triples), with each electrode within the diffusion layer of the other, can be used for studying reactions, in a manner analogous to ring–disk generation–collection and redox recycling experiments (131,132). Unlike with rotating ring–disk electrodes, the product of the reaction at the collector electrode can diffuse back across the narrow gap to the

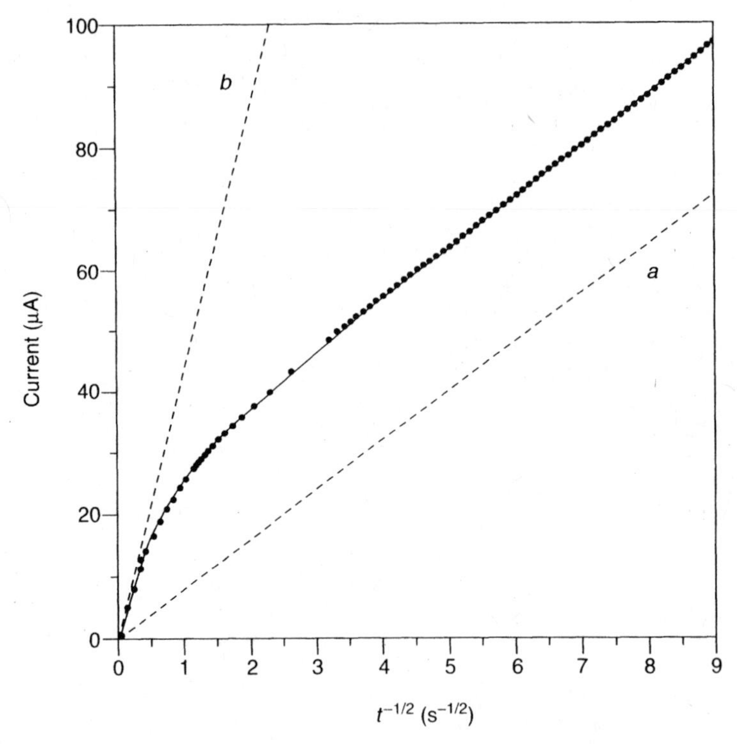

Figure 4.31 Cottrell plot of the chronoamperometric response for 1 × 10^{-3} M $Ru(NH_3)_6^{3+}$ at a Kel-F/gold composite electrode. Points represent experimental data; the solid line indicates the least-squares fit to theory. Dashed lines indicate theoretical Cottrell plots for a macroelectrode with active area equal to the active area of the composite (*a*) and to the geometric area of the composite (*b*). (Reproduced with permission from Ref. 122.)

generator electrode to give higher collection efficiencies. A typical generation–collection experiment at such an array is illustrated in Figure 4.32. Such electrochemical recycling of a reversible redox couple provides a signal amplification (i.e., higher sensitivity) that cannot be achieved at a single electrode, and indicates great promise for various sensing applications (133,134). The coverage of such closely spaced microelectrodes with conducting polymers can form the basis for novel microelectronic (transistor-like) sensing devices (see Chapter 6). The properties and applications of interdigitated arrays of microband electrodes have been reviewed (133). Ordered arrays, featuring regular disk microelectrode sizes and spacing, have potential applications as sensor arrays or high-density DNA and protein chips (discussed in Chapter 6).

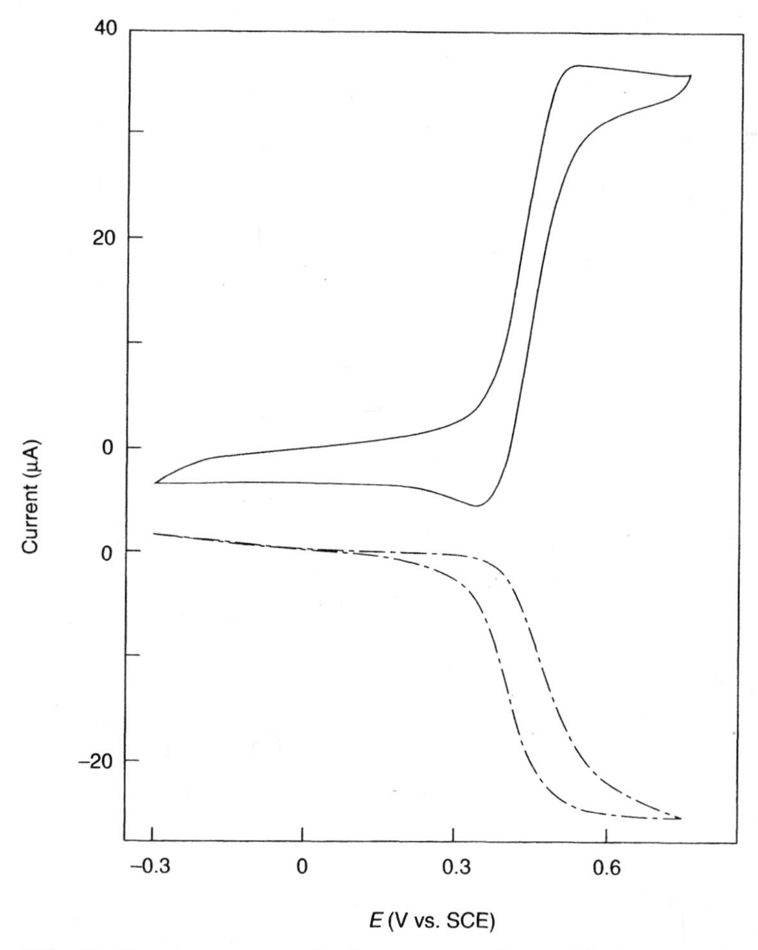

Figure 4.32 Cyclic voltammogram for ferrocene at a 3-μm-wide, 2-μm gap interdigitated microband (solid line). The dotted line represents the current of the collector electrode held at a potential of –0.1 V. (Reproduced with permission from Ref. 132.)

EXAMPLES

Example 4.1 A rotating ring–disk electrode (1600 rpm) yields a disk current of 12.3 μA for the oxidation of a 2×10^{-3} M potassium ferrocyanide solution. Calculate the reduction current observed at the surrounding ring using a 6×10^{-3} M potassium ferrocyanide solution and a rotation speed of 2500 rpm (N = 0.33).

Solution From the Levich equation, (4.5), one can calculate first the disk current under the new conditions:

$$i_D/12.3 = K \times (6 \times 10^{-3}) \times (2500)^{1/2} / K \times (2 \times 10^{-3}) \times (1600)^{1/2}$$
$$i_D = 46.125\,\mu\text{A}$$

Then, from Eq. (4.8), we can solve for the ring current:

$$i_R = -Ni_D = -0.33 \times 12.125 = -15.221\,\mu\text{A}$$

Example 4.2 A rotating mercury film electrode (of 2 mm diameter) yielded a stripping peak current of 2.2 μA for a 1×10^{-8} M lead(II) solution following a 3-min deposition at −1.0 V with a 1600 rpm rotation. Calculate the peak current for a 2.5×10^{-8} M lead(II) solution following a 2-min deposition with a 2500-rpm rotation.

Solution

$$2.2 = K \times (1 \times 10^{-8}) \times 3 \times (1600)^{1/2}$$
$$K = 1.83 \times 10^{6}$$
$$i_p = 1.83 \times 10^{6} (2.5 \times 10^{-8}) \times 2 \times (2500)^{1/2} \qquad i_p = 4.6\,\mu\text{A}$$

PROBLEMS

4.1 What are the advantages of using ultramicroelectrodes for electrochemical measurements?

4.2 Explain how chemically modified electrodes can benefit electrochemical measurements.

4.3 Explain why a carbon composite disk electrode offers improved signal-to-background characteristics compared to a carbon disk electrode of the same geometric area.

4.4 Describe why the oxidation of polypyrrole film results in the uptake of an anion from the surrounding solution.

4.5 Derive the Levich equation for the limiting current at the rotating disk electrode [based on combining Eqs. (3.4) and (1.12)].

4.6 How can you use the rotating–ring disk electrode for detecting short-lived intermediate species?

4.7 Explain why an effective compensation of the ohmic drop is essential for diagnostic applications of cyclic voltammetry (e.g., estimating n from ΔE_p).

4.8 Describe the rationale for using electrodes coated with Nafion films for selective detection of the cationic neurotransmitter dopamine in the presence of coexisting interfering anionic ascorbic acid.

4.9 Explain the reason for including the time-consuming oxygen removal step in pulse polarographic measurements of tin ion in juice samples.

4.10 Explain why and how a change in scan rate affects the shape of the cyclic voltammetric response of an ultramicroelectrodes.

4.11 Propose a modified electrode surface suitable for detecting in situ micromolar concentrations of ferric ion in an industrial stream. What are the challenges for such in situ monitoring?

REFERENCES

1. DeAngelis, T. P.; Bond, R. E.; Brooks, E. E.; Heineman, W. R., *Anal. Chem.* **49**, 1792 (1977).
2. Clark, R.; Ewing, A. G., *Anal. Chem.* **70**, 1119 (1998).
3. Mann, C. K., in *Electroanalytical Chemistry*, A. J. Bard, ed., Marcel Dekker, New York, 1969, Vol. 3, p. 57.
4. Silva, S.; Bond, A. M., *Anal. Chim. Acta* **500**, 307 (2003).
5. Wallace, G. G., *Trends Anal. Chem.* **4**, 145 (1985).
6. Newman, A., *Anal. Chem.* **69**, 369A (1997).
7. Matsue, T.; Aoki, A.; Ando, E.; Uchida, I., *Anal. Chem.* **62**, 409 (1990).
8. Bond, A. M., *Modern Polarographic Methods in Analytical Chemistry*, Marcel Dekker, New York, 1980.
9. He, P.; Avery, J.; Faulkner, L. R., *Anal. Chem.* **54**, 1314 (1982).
10. Bond, A. M.; Luscombe, D.; Tan, S. N.; Walter, F. L., *Electroanalysis* **2**, 195 (1990).
11. Wang, J., *Analyst* **119**, 763 (1994).
12. Erickson, K. A.; Wilding, P., *Clin. Chem.* **39**, 283 (1993).
13. Kemula, W.; Kublik, K., *Anal. Chim. Acta* **18**, 104 (1958).
14. Peterson, W., *Am. Lab.* **12**, 69 (1979).
15. Macchi, G., *J. Electroanal. Chem.* **9**, 290 (1965).
16. Kowalski, Z.; Wang, K.; Osteryoung, R.; Osteryoung, J., *Anal. Chem.* **59**, 2216 (1987).
17. Florence, T. M., *J. Electroanal. Chem.* **27**, 273 (1970).

18. Adams, R. N., *Electrochemistry at Solid Electrodes*, Marcel Dekker, New York, 1969.
19. Engstrom, R. C., *Anal. Chem.* **56**, 890 (1984).
20. Albery, W. J.; Hitchman, M., *Ring-Disk Electrodes*, Clarendon Press, Oxford, UK, 1971.
21. McCreery, R. L., "Carbon electrodes: Structural effects on electron transfer kinetics," in A. J. Bard, ed., *Electroanalytical Chemistry*, Marcel Dekker, New York, 1991, Vol. 18.
22. Chen, P.; McCreery, R. L., *Anal. Chem.* **68**, 3958 (1996).
23. Bokros, J. C., *Carbon* **15**, 355 (1977).
24. Fagan, D.; Hu, I.; Kuwana, T., *Anal. Chem.* **57**, 2759 (1985).
25. Van der Linden, W. E.; Dieker, J. W., *Anal. Chim. Acta* **119**, 1 (1980).
26. Wang, J., *Electrochim. Acta* **26**, 1721 (1981).
27. Olson, C.; Adams, R. N., *Anal. Chim. Acta* **22**, 582 (1960).
28. Urbaniczky, C.; Lundstrom, K., *J. Electroanal. Chem.* **176**, 169 (1984).
29. Rice, M.; Galus, Z.; Adams, R. N., *J. Electroanal. Chem.* **143**, 89 (1983).
30. Kalcher, K., *Electroanalysis* **2**, 419 (1990).
31. Csöregi, E.; Gorton, L.; Marko-Varga, G., *Anal. Chim. Acta* **273**, 59 (1993).
32. Feng, J.; Brazell, M.; Renner, K.; Kasser, R.; Adams, R. N., *Anal. Chem.* **59**, 1863 (1987).
33. Edmonds, T., *Anal. Chim. Acta* **175**, 1 (1985).
34. Compton, R. G.; Foord, J. S.; Marken, F., *Electroanalysis* **15**, 1349 (2003).
35. Xu, J.; Granger, M. C.; Chen, Q.; Lister, T. E.; Strojek, J. W.; Swain, G. M., *Anal. Chem.* **69**, 591A (1997).
36. Terashima, C.; Rao, T. N.; Sarada, B. V.; Tryk, D. A.; Fujishima, A., *Anal. Chem.* **74**, 895 (2002).
37. Marken, F.; Paddon, C. A.; Asogan, D., *Electrochem. Commun.* **4**, 62 (2002).
38. Kozlowska, H.; Conway, B.; Sharp, W., *J. Electroanal. Chem.* **43**, 9 (1973).
39. Johnson, D. C.; LaCourse, W., *Anal. Chem.* **62**, 589A (1990).
40. Colon, L.; Dadoo, R.; Zare, R., *Anal. Chem.* **65**, 476 (1993).
41. Cassela, I. G.; Cataldi, T. R.; Salvi, A.; Desimoni, E., *Anal. Chem.* **65**, 3143 (1993).
42. Wang, J.; Lu, J.; Hocevar, S.; Farias, P.; Ogorevc, B., *Anal. Chem.* **72**, 3218 (2000).
43. Wang, J., *Electroanalysis* **17**, 1341 (2005).
44. Wasmus, S.; Kuver, A., *J. Electroanal. Chem.* **461**, 14 (1999).
45. Murray, R. W.; Ewing, A. G.; Durst, R. A., *Anal. Chem.* **59**, 379A (1987).
46. Wang, J., *Electroanalysis* **3**, 255 (1991).
47. Baldwin, R. P.; Thomsen, K. N., *Talanta* **38**, 1 (1991).
48. Zhong, C.; Porter, M. D., *Anal. Chem.* **67**, 709A (1995).
49. Bain, C.; Whitsides, G., *Angew Chem. Int. Ed. Eng.* **28**(4), 506 (1989).
50. Mandler, D.; Turyan, I., *Electroanalysis* **8**, 207 (1996).
51. Piner, R. D.; Zhu, J.; Zu, F.; Hong, S.; Mirkin, C. A., *Science* **283**, 661 (1999).
52. Kumar, A.; Whitesides, G. M., *Appl. Phys. Lett.* **63**, 2002 (1993).
53. Armstrong, F. A.; Wilson, G. S., *Electrochim. Acta* **45**, 2623 (2000).

54. Baughman, R. H.; Zakhidov, A.; de Heer, W. A., *Science* **297**, 787 (2002).
55. Wang, J.; Musameh, M., *Anal. Chem.* **75**, 2075 (2003).
56. Banks, C. E.; Moore, R. R.; Compton, R. G., *Chem. Commun.* **16**, 1804 (2004).
57. Gooding, J. J.; Wibowo, R.; Liu, J. Q.; Yang, W.; Losic, D.; Orbons, S.; Mearns, F. J.; Shapter, J. G.; Hibbert, D. B., *J. Am. Chem. Soc.* ***125***, 9006 (2003).
58. Patolsky, F.; Weizmann, Y.; Willner, I., *Angew Chem. Int. Ed.* **43**, 2113 (2004).
59. Wang, J., *Electroanalysis* **17**, 7 (2005).
60. Lev, O.; Tsionsky, M.; Rabinovich, I.; Glezer, V.; Sampath, S.; Pankratov, I.; Gun, J., *Anal. Chem.* **67**, 22A (1999).
61. Petit-Dominquez, M.; Shen, H.; Heineman, W. R.; Seliskar, C. J., *Anal. Chem.* **69**, 703 (1997).
62. Tsionsky, M.; Gun, G.; Glezer, V.; Lev, O., *Anal. Chem.* **66**, 1747 (1994).
63. Wang, J.; Pamidi, P., *Anal. Chem.* **69**, 4490 (1997).
64. Marx, S.; Zaltsman, A.; Turyan, I.; Mandler, D., *Anal. Chem.* **76**, 120 (2004).
65. Zak, J.; Kuwana, T., *J. Electroanal. Chem.* **150**, 645 (1983).
66. Cox, J.; Jaworski, R.; Kulesza, P., *Electroanalysis* **3**, 869 (1991).
67. Gorton, L., *J. Chem. Soc. Faraday Trans.* **82**, 1245 (1986).
68. Halbert, M.; Baldwin, R., *Anal. Chem.* **57**, 591 (1985).
69. Malinski, T.; Taha, Z., *Nature* **358**, 676 (1992).
70. Wang, J.; Taha, Z., *Anal. Chem.* **62**, 1413 (1990).
71. Karayakin, K., *Electroanalysis* **13**, 813 (2001).
72. Carrette, L.; Friedrich, K. A.; Stimming, U., *Fuel Cells* **1**, 5 (2001).
73. Wang, J., "Voltammetry after nonelectrolytic preconcentration," in A. J. Bard, ed., *Electroanalytical Chemistry*, Marcel Dekker, New York, 1989, Vol. 16, p. 1.
74. Guadalupe, A.; Abruna, H., *Anal. Chem.* **57**, 142 (1985).
75. Arrigan, D. W., *Analyst* **119**, 1953 (1994).
76. Ugo, P.; Moreto, L. M., *Electroanalysis* **7**, 1105 (1995).
77. Prabhu, S.; Baldwin, R.; Kryger, L., *Anal. Chem.* **59**, 1074 (1987).
78. Wang, J.; Martinez, T., *Electroanalysis* **1**, 167 (1989).
79. Oyama, N.; Anson, F. C., *Anal. Chem.* **52**, 1192 (1980).
80. Wang, J.; Lu, Z., *J. Electroanal. Chem.* **266**, 287 (1989).
81. Prabhu, S.; Baldwin, R.; Kryger, L., *Electroanalysis* **1**, 13 (1989).
82. Ugo, P.; Ballarin, B.; Daniele, S.; Mazzocchin, G., *Anal. Chim. Acta* **244**, 29 (1991).
83. Turyan, I.; Mandler, D., *Anal. Chem.* **66**, 58 (1994).
84. Yang, W. R.; Jaramillo, D.; Gooding, J. J.; Hibbert, D. B.; Zhang, R.; Willett, G. D.; Fisher, K., *J. Chem. Commun.* 1982 (2001).
85. Malinski, T.; Ciszewski, A.; Fish, J.; Czuchajowski, L., *Anal. Chem.* **62**, 90 (1990).
86. Sasso, S. V.; Pierce, R.; Walla, R.; Yacynych. A., *Anal. Chem.* **62**, 1111 (1990).
87. Nagy, G.; Gerhardt, G.; Oke, A.; Rice, M.; Adams, R. N., *J. Electroanal. Chem.* **188**, 85 (1985).
88. Wang, J.; Lu, Z., *Anal. Chem.* **62**, 826 (1990).
89. Wang, J.; Wu, H.; Angnes, S., *Anal. Chem.* **65**, 1893 (1993).
90. Wang, J.; Tuzhi, P., *Anal. Chem.* **58**, 3257 (1986).

91. Sittampalam, G.; Wilson, G., *Anal. Chem.* **55**, 1608 (1983).
92. Ohsaka, T.; Hirokawa, T.; Miyamoto, H.; Oyama, N., *Anal. Chem.* **59**, 1758 (1987).
93. Wang, J.; Golden, T., *Anal. Chem.* **61**, 1397 (1989).
94. Cheng, Q.; Brajter-Toth, A., *Anal. Chem.* **64**, 1998 (1992).
95. Bidan, G., *Sensors and Actuators B* **6**, 45 (1992).
96. Kanatzidis, M., *Chem. Eng. News* 36 (Dec. 3, 1990).
97. Ikariyama, T.; Heineman, W. R., *Anal. Chem.* **58**, 1803 (1986).
98. Ivaska, A., *Electroanalysis* **3**, 247 (1991).
99. Teasdale, P.; Wallace, G., *Analyst* **118**, 329 (1993).
100. Liu, H.; Kameoka, J.; Caplewski, D. A.; Craighead, H. G., *Nano. Lett.* **4**, 671 (2004).
101. Martin, C. R., *Science* **266**, 1961 (1994).
102. Ramanathan, K.; Bangar, M. A.; Yun, M.; Chen, W.; Mulchandani, A.; Myang, N. V. *Nano. Lett.* **4**, 1237 (2004).
103. Piletsky, S. A.; Turner, A. P., *Electroanalysis* **14**, 317 (2002).
104. Malitesta, C.; Losito, I.; Zambonin, P. G., *Anal. Chem.* **71**, 1366 (1999).
105. Wightman, R. M., *Science* **240**, 415 (1988).
106. Wightman, R. M., *Anal. Chem.* **53**, 1125A (1981).
107. Bond, A. M., *Analyst* **119**, R1 (1994).
108. Zoski, C. G., *Electroanalysis* **14**, 1041 (2002).
109. Kennedy, R.; Huang, L.; Atkinson, M.; Dush, P., *Anal. Chem.* **65**, 1882 (1993).
110. Cahill, P. S.; Walker, Q. D.; Finnegan, J. M.; Mickelson, G. E.; Travis, E. R.; Wightman, R. M., *Anal. Chem.* **68**, 3180 (1996).
111. Bedioui, F.; Villeneuve, N., *Electroanalysis* **15**, 5 (2003).
112. Fan, F.; Kwak, J.; Bard, A. J., *J. Am. Chem. Soc.* **118**, 9669 (1996).
113. Anderson, B. B.; Ewing, A. G., *J. Pharm. Biomed. Anal.* **19**, 15 (1999).
114. Andrieux, C. P.; Hapiot, P.; Saveant, J. M., *Electroanalysis* **2**, 183 (1990).
115. White, R. J.; White, H. S., *Anal. Chem.* **77**, 215A (2005).
116. Heinze, J., *Angew. Chem. (Engl. Ed.)* **32**, 1268 (1993).
117. Penner, R.; Lewis, N., *Chem. Ind.* 788 (Nov. 4, 1991).
118. Watkins, J.; Chen, J.; White, H.; Abruna, H.; Maisonhaute, E.; Amatore, C., *Anal. Chem.* **75**, 3962 (2003).
119. Watkins, J. J.; Zhang, B.; White, H. S., *J. Chem. Educ.* **82**, 712 (2005).
120. Arrigan, D. W. M., *Analyst* **129**, 1157 (2004).
121. Shao, Y.; Mirkin, M. V.; Fish, G.; Kokotov, S.; Palankar, D.; Lewis, A., *Anal. Chem.* **69**, 1627 (1997).
122. Tallman, D. E.; Petersen, S. L., *Electroanalysis* **2**, 499 (1990).
123. Thormann, W.; van den Bosch, P.; Bond, A. M., *Anal. Chem.* **57**, 2764 (1985).
124. Feeny, R.; Kounaves, S. P., *Electroanalysis* **12**, 677 (2000).
125. Weisshaar, D. E.; Tallman, D. E., *Anal. Chem.* **55**, 1146 (1983).
126. Wang, J.; Brennsteiner, A.; Sylwester, A. P., *Anal. Chem.* **62**, 1102 (1990).
127. Deutscher, R.; Fletcher, S., *J. Electroanal. Chem.* **239**, 17 (1988).
128. Penner, R. M.; Martin, C. R., *Anal. Chem.* **59**, 2625 (1987).

129. Hulteen, J. C.; Martin, C. R., *J. Mater. Chem.* **7**, 1075 (1997).
130. Gueshi, T.; Tokuda, K.; Matsuda, H., *J. Electroanal. Chem.* **89**, 247 (1978).
131. Bard, A. J.; Crayton, J.; Kittlesen, G.; Shea, T.; Wrighton, M. S., *Anal. Chem.* **58**, 2321 (1986).
132. Niwa, O.; Morita, M.; Tabei, H., *Anal. Chem.* **62**, 447 (1990).
133. Niwa, O., *Electroanalysis* **7**, 606 (1995).
134. Thomas, J.; Kim, S.; Hesketh, P.; Halsall, H. B.; Heineman, W. R., *Anal. Chem.* **76**, 2700 (2004).

5

POTENTIOMETRY

5.1 PRINCIPLES OF POTENTIOMETRIC MEASUREMENTS

In potentiometry, information on the composition of a sample is obtained through the potential appearing between two electrodes. Potentiometry is a classical analytical technique with roots before the twentieth century. However, the rapid development of new selective electrodes and more sensitive and stable electronic components since 1970 has tremendously expanded the range of analytical applications of potentiometric measurements. Selective potentiometric electrodes are currently widely used in many fields, including clinical diagnostics, industrial process control, environmental monitoring, and physiology. For example, such devices are used in nearly all hospitals around the globe for assessing several physiologically important blood electrolytes (K^+, Na^+, Ca^{2+}, Mg^{2+}, H^+, Cl^-) relevant to various health problems. The speed at which this field has developed is a measure of the degree to which potentiometric measurements meet the needs of the analytical chemist for rapid, low-cost, and accurate analysis. In this chapter, the principles of direct potentiometric measurements, based on ion-selective electrodes (ISEs), will be described. ISEs are chemical sensors with the longest history. The field of ISE bridges fundamental membrane science with fundamental host–guest chemistry. (The second major part of potentiometry, the so-called potentiometric titrations, will not be covered.) General books devoted exclusively to direct potentiometry can be found in Refs. 1–5.

The equipment required for direct potentiometric measurements includes an ion-selective electrode, a reference electrode, and a potential-measuring device (a pH/millivolt meter that can read 0.2 mV or better) (Fig. 5.1). Conventional voltmeters cannot be used because only very small currents can be drawn. The reference electrode should provide a highly stable potential for an extended period of time. The ion-selective electrode is an indicator electrode capable of selectively measuring the activity of a particular ionic species (known as the *primary* or *analyte ion*). Such electrodes exhibit a fast response and a wide linear range, are not affected by color or turbidity, are not destructive, and are very inexpensive. Ion-selective electrodes can be assembled conveniently in a variety of shapes and sizes. Specially designed cells allow flow or microliter analyses (see, e.g., Section 5-3).

Ion-selective electrodes are mainly membrane-based devices, consisting of permselective ion-conducting materials, which separate the sample from the inside of the electrode (Fig. 5.2). On the inside is a filling solution containing the ion of interest at a constant activity. The membrane is usually nonporous, water insoluble, and mechanically stable. The composition of the membrane is designed to yield a potential that is primarily due to the ion of interest (via selective binding processes, e.g., ion exchange, which occur at the membrane–solution interface). The trick is to find a membrane that will selectively bind the analyte ions, leaving co-ions behind. Membrane materials, possessing different ion recognition properties, have thus been developed to impart high

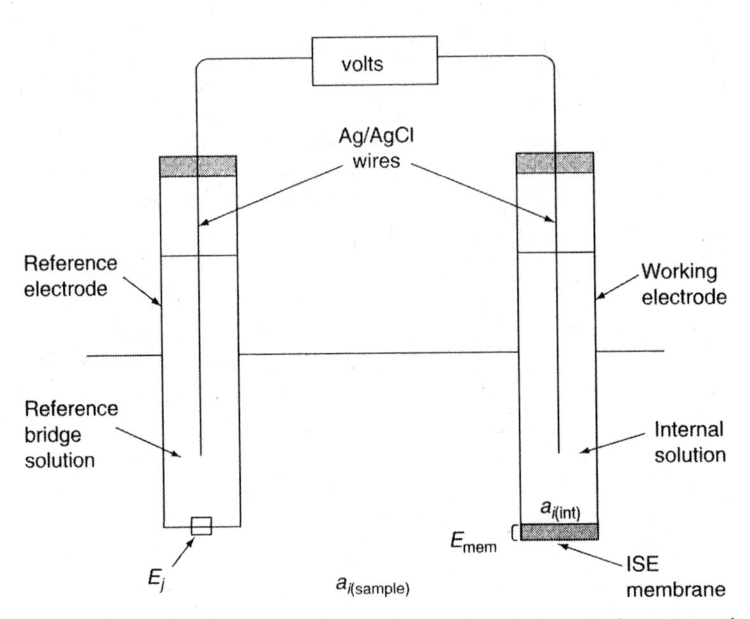

Figure 5.1 Schematic diagram of an electrochemical cell for potentiometric measurements.

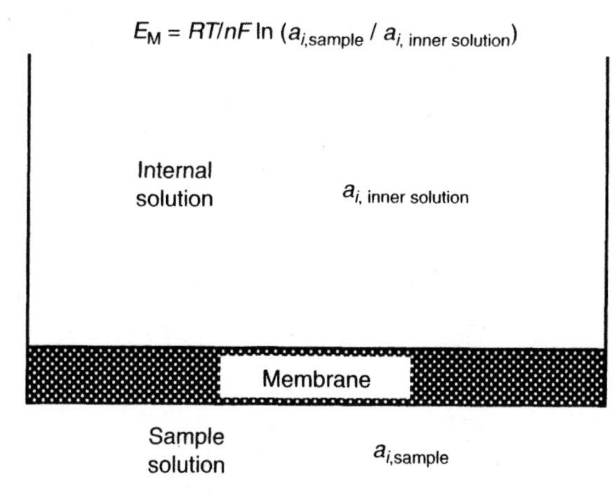

Figure 5.2 Membrane potential reflects the gradient of activity of the analyte ion in the inner and outer (sample) solutions.

selectivity (see Section 5.2). Detailed theory of the processes at the interface of these membranes, which generate the potential, is available elsewhere (6–8). The ion recognition (binding) event generates a phase boundary potential at the membrane–sample interface:

$$E_{PB} = \frac{RT}{z_i F} \ln k_i + \frac{RT}{z_i F} \ln \frac{a_i(\text{aq})}{a_i(\text{org})} \tag{5.1}$$

where R is the universal gas constant (8.134 J K^{-1} mol^{-1}), F is the Faraday constant, and T is the absolute temperature; a_i(aq) and a_i(org) are the activities of the primary ion (with charge z_i) in the aqueous sample and the contacting organic phase boundary, respectively, and k_i is a function of the relative free energies of solvation in both the sample and the membrane phase ($k_i = \exp(\{\mu_i^\circ(\text{aq}) - \mu_i^\circ(\text{org})\})/RT$, where μ_i°(aq) and μ_i°(org) are the chemical standard potentials of the ion I^{z_i+} in the respective phase). The first term on the right-hand side of Eq. (5.1) is in fact the standard potential, which is constant for a given ion but varies from ion to ion. The phase boundary potential is a consequence the unequal distribution of the analyte ions across the boundary. From Eq. (5.1) it is apparent that a selective binding to a cation in the membrane decreases its activity in the membrane phase and thus increases the phase boundary potential.

Another phase boundary potential is developed at the inner surface of the membrane (at the membrane/filling solution interface). The membrane potential corresponds to the potential difference across the membrane:

$$E = \frac{RT}{nF} \ln(a_{i,\text{sample}} / a_{i,\text{int soln}}) \tag{5.2}$$

The potential of the ion-selective electrode is generally monitored relative to the potential of a reference electrode. Since the potential of the reference electrode is fixed, and the activity of the ion in the inner solution is constant, the measured cell potential reflects the potential of the ISE, and can thus be related to the activity of the target ion in the sample solution. Ideally, the response of the ISE should obey the following equation

$$E = K + (2.303RT/z_iF)\log a_i \tag{5.3}$$

where E is the potential, and z_i and a_i are the ionic charge and activity, respectively, of the ion. The constant K includes all sample-independent potential contributions, which depends on various factors (influenced by the specific design of the ISE). Equation (5.3) predicts that the electrode potential is proportional to the logarithm of the activity of the ion monitored. For example, at room temperature a 59.1-mV change in the electrode potential should result from a 10-fold change in the activity of a monovalent ion (z = 1). Similar changes in the activity of a divalent ion should result in a 29.6-mV change of the potential. A 1-mV change in the potential corresponds to 4% and 8% changes in the activity of monovalent and divalent ions, respectively. The term "Nernstian behavior" is used to characterize such behavior. In contrast, when the slope of the electrode response is significantly smaller than $59.1/z_i$, the electrode is characterized by a sub-Nernstian behavior.

It should be noted again that ISEs sense the activity, rather than the concentration of ions in solution. The term "activity" is used to denote the effective (active) concentration of the ion. The difference between concentration and activity arises because of ionic interactions (with oppositely charged ions) that reduce the effective concentration of the ion. The activity of an ion i in solution is related to its concentration c_i by the following equation:

$$a_i = f_i c_i \tag{5.4}$$

where f_i is the activity coefficient. The activity coefficient depends on the types of ions present and on the total ionic strength of the solution. The activity coefficient is given by the Debye–Hückel equation

$$\log f_i = \frac{-0.51 z_i^2 \sqrt{\mu}}{1 + \sqrt{\mu}} \quad \text{(at 25°C)} \tag{5.5}$$

where μ is the ionic strength. The ionic strength refers to the concentration of all ions in the solution and also takes into account their charge. The activity coefficient thus approaches unity (i.e., $a_i \cong C_i$) in very dilute solutions. The departure from unity increases as the charge of the ion increases.

Equation (5.3) has been written on the assumption that the electrode responds only to the ion of interest, i. In practice, no electrode responds exclusively to the ion specified. The actual response of the electrode in a binary mixture of the primary and interfering ions (i and j, respectively) is given by the *Nikolskii–Eisenman equation* (9):

$$E = K + (2.303RT/z_i F)\log\left(a_i + k_{ij}a_j^{z_i/z_j}\right) \tag{5.6}$$

where k_{ij} is the selectivity coefficient, a quantitative measure of the electrode ability to discriminate against the interfering ion (i.e., a measure of the relative affinity of ions i and j toward the ion-selective membrane). For example, if an electrode is 50 times more responsive to i than to j, k_{ij} has a value of 0.02. A k_{ij} value of 1.0 corresponds to a similar response for both ions. When $k_{ij} \gg 1$, then the ISE responds better to the interfering ion j than to the target ion i. Usually, k_{ij} is smaller than 1, which means that the ISE responds more selectively to the target ion. The lower the value of k_{ij}, the more selective is the electrode. Selectivity coefficients lower than 10^{-5} have been achieved for several electrodes. For an ideally selective electrode, the k_{ij} would equal zero (i.e., no interference). Obviously, the error in the activity a_i due to the interference of j would depend on their relative levels. The term z_i/z_j corrects for a possible charge difference between the target and interfering ions. Normally, the most serious interferences have the same charge as the primary ion so that $z_i/z_j = 1$. In practice, the contribution of all interfering ions present in the sample matrix ($\Sigma k_{ij}a^{z_i/z_j}$) should be included in the Nikolskii–Eisenman equation. For example, for a sodium electrode immersed in a mixture of sodium, potassium, and lithium, the response is given by

$$E = K + 2.303RT/F\log(a_{\mathrm{Na}} + k_{\mathrm{Na,K}}a_{\mathrm{K}} + k_{\mathrm{Na,Li}}a_{\mathrm{Li}}) \tag{5.7}$$

Accordingly, an ISE displays a selective response when the activity of the primary ion is much larger than the summation term of the interferents, specifically, $a_i \gg \Sigma k_{ij}a_j^{z_i/z_j}$. Under this condition, the effect of interfering ions is negligible, and changes in the measured potential can be related with confidence to variations in the activity of the target ion. The selectivity coefficients thus serve as guidelines as to how far a given ISE should be applicable for a particular analytical problem. Nonselective ISEs are rarely useful for real-life applications (with the exception of their combination with the operation of ISE arrays; see Section 6.4). In reality, equations with more than two components are rarely used. Deviations from the Nikolski–Eisenman equation have been reported for various situations (particularly for mixtures of ions of different charge, in the case of non-Nernstian behavior of interfering ions, and due to the concentration dependence of k_{ij}).

It is important for the analytical chemist to realize the selectivity coefficient of a particular electrode. Various methods have been suggested for determining the selectivity coefficient, including the fixed-interference method, separate solution method, and the fixed primary ion method (10,11). The most popular fixed interference method involves two solutions, one containing a constant concentration of the interfering ion and the second, containing a zero concentration. Also popular is the separate solution method, which involves the preparation of calibration curves for each ion. As selectivity is a complex function of the membrane composition and the experimental design, the values of selectivity coefficients should be regarded as operationally defined (i.e., valid for the particular set of conditions used for their determination).

Usually, the analytical chemist needs to determine the concentration of the ion of interest rather than its activity. The obvious approach to converting potentiometric measurements from activity to concentration is to make use of an empirical calibration curve, such as the one shown in Figure 5.3. Electrodes potentials of standard solutions are thus measured and plotted (on a semilog paper) versus the concentration. Since the ionic strength of the sample is seldom known, it is often useful to add a high concentration of an electrolyte to the standards and the sample to maintain approximately the same ionic strength (i.e., the same activity coefficient). The ionic strength adjustor is usually a buffer (since pH control is also desired for most ISEs). The empirical calibration plot thus yields results in terms of concentration. Theoretically,

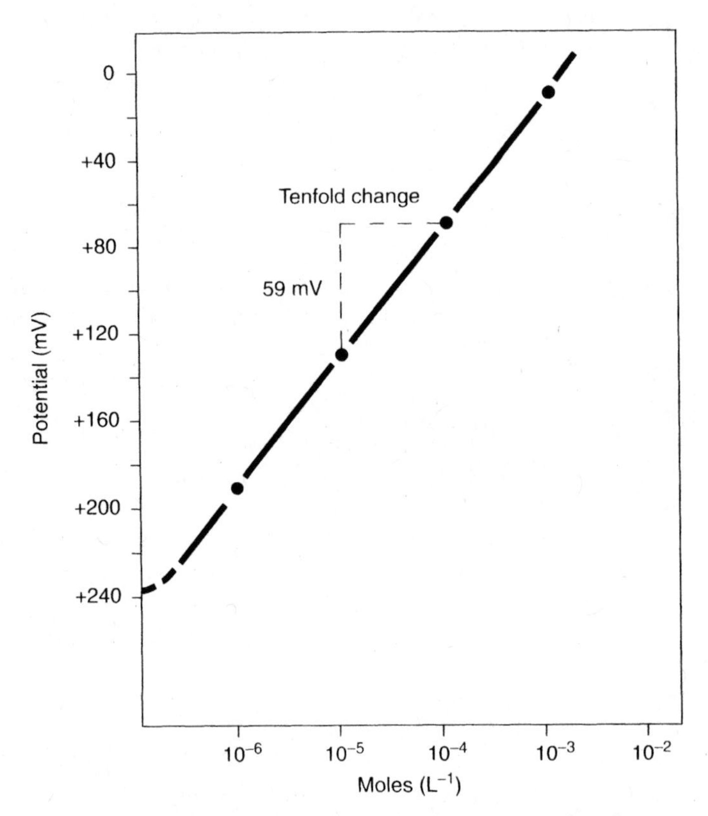

Figure 5.3 Typical calibration plot for a monovalent ion.

such a plot should yield a straight line, with a slope of approximately $59/z_i$ mV (Nernstian slope). Detection by means of ion-selective electrodes may be performed over an exceedingly broad concentration range, which, for certain electrodes, may embrace five orders of magnitude. In practice, the usable range depends on other ions in the solution. Departure from the linearity is commonly observed at low concentrations (about 10^{-6} M) due to the presence of coexisting ions [Eq. (5.6)]. The extent of such departure (and the minimum activity that can be accurately measured) depend on the selectivity coefficient as well as upon the level of the interfering ion (Fig. 5.4). The detection limit for the analyte ion is defined by

$$a_{i,\min} = k_{ij} a_j^{z_i/z_j} \tag{5.8}$$

and corresponds to the activity of i at the intersection of the asymptotes in the $E/\log a_i$ calibration plot, that is, where the extrapolated linear and zero-slope segments meet (see Ref. 12 and Fig 5.5). It is only when the plot becomes almost horizontal that the activity measurement becomes impossible. At high concentrations of the ions of interest, interference by species of opposite charge [not described by Eq. (5.6)] may lead to deviation from the linear electrode response.

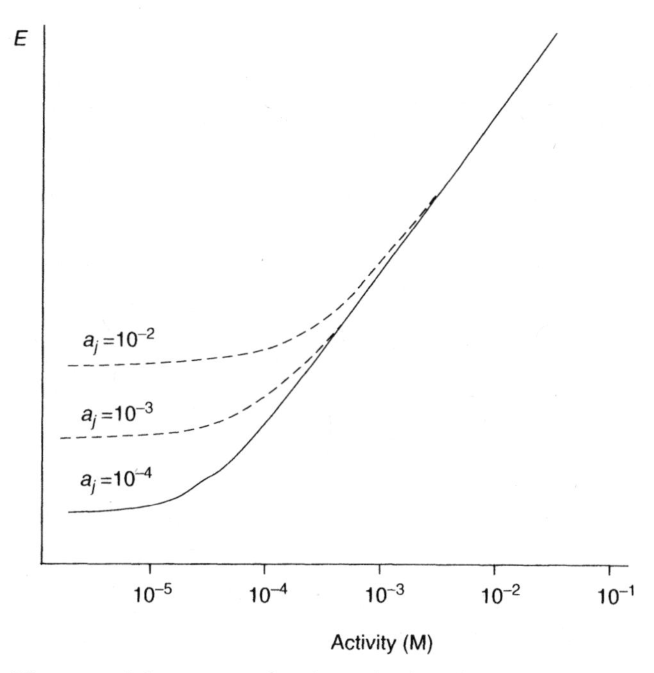

Figure 5.4 The potential response of an ion-selective electrode versus activity of ion i in the presence of different levels of an interfering ion j.

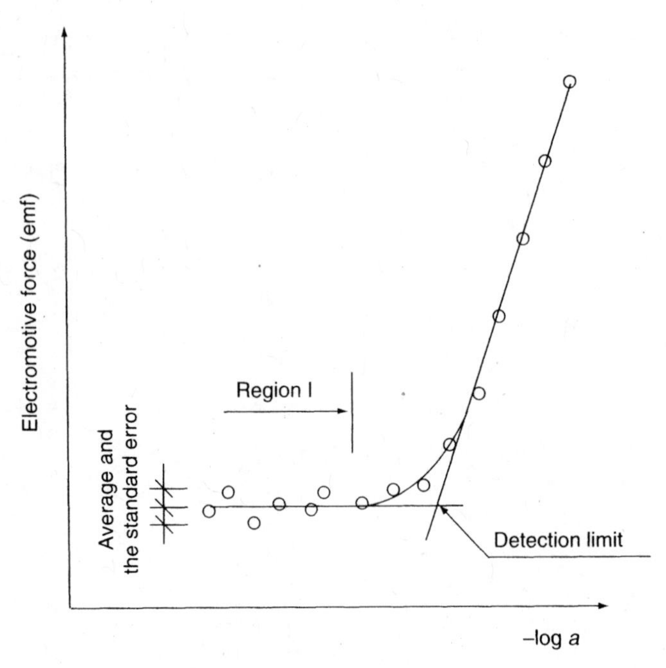

Figure 5.5 Determination of the detection limit of ion-selective electrodes. (Reproduced with permission from Ref. 12.)

The logarithmic response of ISEs can cause major accuracy problems. Very small uncertainties in the measured cell potential can thus cause large errors. (Recall that an uncertainty of ±1 mV corresponds to a relative error of ~4% in the concentration of a monovalent ion.) Since potential measurements are seldom better than 0.1 mV uncertainty, best measurements of monovalent ions are limited to about 0.4% relative concentration error. In many practical situations, the error is significantly larger. The main source of error in potentiometric measurements is actually not the ISE, but rather changes in the reference electrode junction potential, namely, the potential difference generated between the reference electrolyte and sample solution. The junction potential is caused by an unequal distribution of anions and cations across the boundary between two dissimilar electrolyte solutions (which results in ion movement at different rates). When the two solutions differ only in the electrolyte concentration, such liquid junction potential is proportional to the difference in transference numbers of the positive and negative ions and to the log of the ratio of the ions on both sides of the junction:

$$E = \frac{RT}{F}(t_1 - t_2)\ln\frac{a_i(1)}{a_i(2)} \tag{5.9}$$

Changes in the reference electrode junction potential result from differences in the composition of the sample and standard solutions (e.g., on switching from whole blood samples to aqueous calibrants). One approach to alleviate this problem is to use an intermediate salt bridge, with a solution (in the bridge) of ions of nearly equal mobility (e.g., concentrated KCl). Standard solutions with an electrolyte composition similar to the sample are also desirable. These precautions, however, will not eliminate the problem completely. Other approaches to address this and other changes in the cell constant have been reviewed (13).

5.2 ION-SELECTIVE ELECTRODES

The discussion in Section 5.1 clearly illustrates that the most important response characteristic of an ISE is selectivity. Depending on the nature of the membrane material used to impart the desired selectivity, ISEs can be divided into three groups: glass, liquid, or solid electrodes. More than three dozen ISEs are commercially available and are widely used (although many more have been reported in the literature). Such electrodes are produced by firms such as Thermo-Electron (Orion), Radiometer, Corning Glass, Beckman, Hitachi, or Sensorex. Recent research activity has led to exciting advances in the area of ISE, including dramatic lowering of their detection limits (to enable trace analysis), identification of new ionophore systems, or new membranes responding to important polyionic species (e.g., heparin) or to neutral species (such as surfactants) (14).

5.2.1 Glass Electrodes

Glass electrodes are responsive to univalent cations. The selectivity for these cations is achieved by varying the composition of a thin ion-sensitive glass membrane.

5.2.1.1 pH Electrodes The most common potentiometric device is the pH electrode. This electrode has been widely used for pH measurements for several decades. Besides direct pH measurements, the pH glass electrode is commonly employed as the transducer in various gas and biocatalytic sensors, involving proton-generating/consuming reactions (see Chapter 6). Its remarkable success is attributed to its outstanding analytical performance, in particular its extremely high selectivity for hydrogen ions, its remarkably broad response range, and its fast and stable response. The phenomenon of glass selectivity was reported by Cremer in 1906 (15). Glass pH electrodes of different configurations and dimensions have been in routine use since the early 1940s following their commercial introduction by A. Beckman. A schematic of a commonly used configuration is shown in Figure 5.6. This consists of a

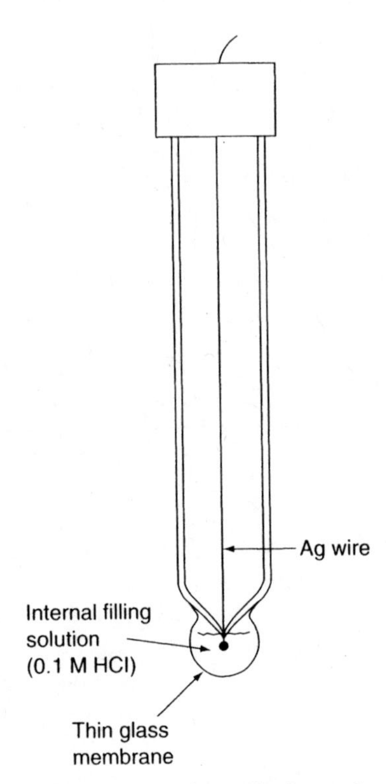

Figure 5.6 A glass pH electrode.

thin, pH-sensitive glass membrane sealed to the bottom of an ordinary glass tube. The composition of the glass membrane is carefully controlled. Usually, it consists of a three-dimensional silicate network, with negatively charged oxygen atoms, available for coordinating cations of suitable size. Some of the more popular glasses have three-component compositions of 72% SiO_2–22% Na_2O–6% CaO or 80% SiO_2–10% Li_2O–10% CaO. Inside the glass bulb are a dilute hydrochloric acid solution and a silver wire coated with a layer of silver chloride. The electrode is immersed in the solution whose pH is to be measured, and connected to an external reference electrode. (In the so-called combination electrode, the external reference electrode is combined with the ion-selective electrode into one body.) The rapid equilibrium established across the glass membrane, with respect to the hydrogen ions in the inner and outer solutions, produces a potential:

$$E = K + (RT/F)\ln(\mathrm{H}^+)_{\mathrm{inner}}/(\mathrm{H}^+)_{\mathrm{outer}} \tag{5.10}$$

The potential of the electrode is registered with respect to the external reference electrode. Hence, the cell potential (at 25°C and after introducing the definition of pH) follows the relation

$$E_{cell} = K' + 0.059\ \text{pH} \tag{5.11}$$

The measured potential is thus a linear function of pH; an extremely wide (10–14 decades) linear range is obtained, with calibration plots yielding a slope of 59 mV/pH unit. The overall mechanism of the response is complex. The selective response is attributed to the ion exchange properties of the glass surface, in particular replacement of sodium ions associated with the silicate groups in the glass by protons:

$$\text{Na}^+_{\text{glass}} + \text{H}_3\text{O}^+_{\text{soln}} = \text{H}_3\text{O}^+_{\text{glass}} + \text{Na}^+_{\text{soln}} \tag{5.12}$$

The theory of the response mechanism has been thoroughly discussed (16).

The user must be alert to some shortcomings of the glass pH electrode. For example, in solutions of pH ≥ 11, the electrode shows a so-called alkaline error in which it also responds to changes in the level of alkali metal ions (particularly sodium):

$$E_{cell} = K + 0.059 \log([\text{H}_3\text{O}^+] + k_{\text{H,Na}}[\text{Na}^+]) \tag{5.13}$$

As a result, the pH is lower than the true value (Fig. 5.7). This error is greatly reduced if the sodium oxide in the glass is replaced by lithium oxide. Still, even

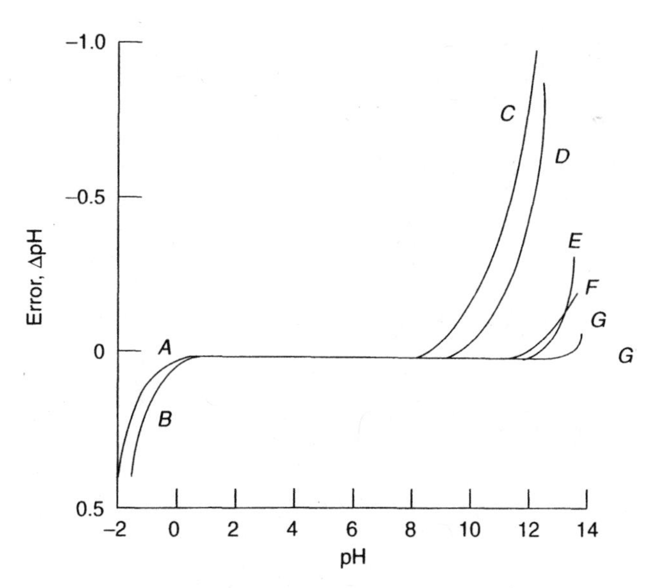

Figure 5.7 The alkaline and acid errors of several glass pH electrodes: *A*, Corning 015/H_2SO_4; *B*, Corning 015/HCl; *C*, Corning 015/1 M Na^+; *D*, Beckman-GP/1 M Na^+; *E*, L&N BlackDot/1 M Na^+; *F*, Beckman E/1 M Na^+; *G*, Ross electrode. (Reproduced with permission from Ref. 17.)

with new glass formulations (with $k_{H,Na} < 10^{-10}$), errors can be appreciable when measurements are carried out in highly basic solutions (e.g., NaOH). Many glass electrodes also exhibit erroneous results in highly acidic solutions (pH < 0.5); the so-called acid error yields higher pH readings than the true value (Fig. 5.7).

Before using the pH electrode, it should be calibrated using two (or more) buffers of known pH. Many standard buffers are commercially available, with an accuracy of ±0.01 pH unit. Calibration must be performed at the same temperature at which the measurement will be made; care must be taken to match the temperature of samples and standards. The exact procedure depends on the model of pH meter used. Modern pH meters, such as the one shown in Figure 5.8, are microcomputer-controlled, and allow double-point calibration, slope calculation, temperature adjustment, and accuracy to ±0.001 pH unit, all with few basic steps. The electrode must be stored in an aqueous solution when not in use, so that the hydrated gel layer of the glass does not dry out. A highly stable response can thus be obtained over long time periods. As with other ion-selective electrodes, the operator should consult the manufacturer's instructions for proper use. Commercial glass electrodes are remarkably

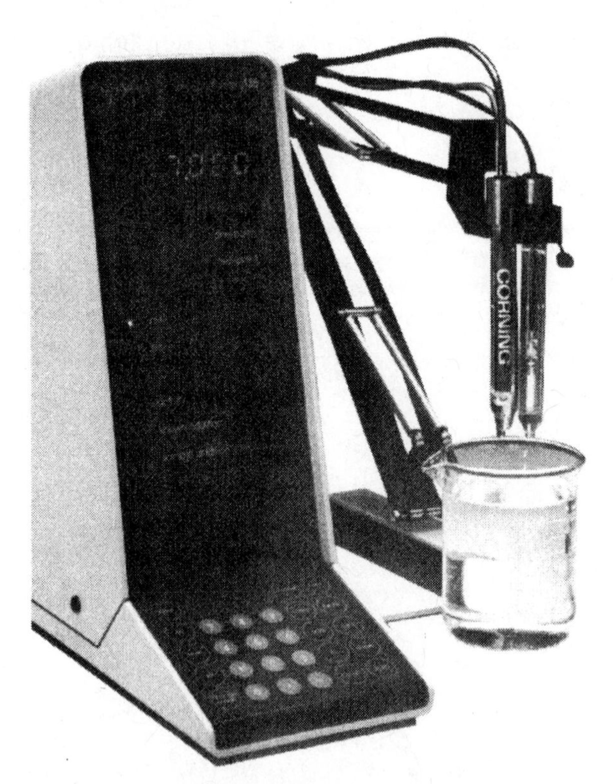

Figure 5.8 A modern microprocessor-controlled pH meter.

robust and, with proper care, will last for more than a year. Proper maintenance of the reference electrode is also essential to minimize errors.

Measurements of pH can also be performed using other types of potentiometric sensors. Nonglass electrodes offer various advantages for certain pH measurements (particularly intravascular and intraluminal clinical applications, food assays, and operation in fluoride media), including ease of preparation, low electrical resistance, and safety in handling. The most common examples are the quinhydrone electrode in which the response is due to a proton transfer redox reaction (of the quinone–hydroquinone couple) and the antimony electrode (based on the redox reaction between antimony and antimony oxide involving protons). Other metal—metal oxide couples, such as palladium—palladium oxide, have been applied for pH measurements. Membrane electrodes based on various neutral hydrogen ion carriers (e.g., tridodecylamine) can also be employed (18). The resulting electrodes exhibit excellent selectivity, reproducibility, and accuracy, but their dynamic range is inferior compared with glass electrodes. (Such a range appears to depend on the acidity constant of the incorporated ionophore.) New pH sensors based on new glass compositions or nonglass formulations are currently (as of 2005) being developed in various laboratories. While such electrodes may be useful for specific applications, glass electrodes are likely to remain the choice for routine analytical applications.

5.2.1.2 Glass Electrodes for Other Cations From the early days of glass pH electrodes, alkaline solutions were noted to display some interference on the pH response. Deliberate changes in the chemical composition of the glass membrane (along with replacements of the internal filling solution) have thus led to electrodes responsive to monovalent cations other than hydrogen, including sodium, ammonium, and potassium (16). This usually involves the addition of B_2O_3 or Al_2O_3 to sodium silicate glasses, to produce anionic sites of appropriate charge and geometry on the outer layer of the glass surface. For example, the sodium- and ammonium-selective glasses have the compositions 11%Na_2O–18%Al_2O_3–71%SiO_2 and 27%Na_2O–4%Al_2O_3–69%SiO_2, respectively. Unlike sodium silicate glasses (used for pH measurements), these sodium aluminosilicate glasses possess what may be termed $AlOSiO^-$ sites with a weaker electrostatic field strength and a marked preference for cations other than protons. The overall mechanism of the electrode response is complex but involves a combination of surface ion exchange and ion diffusion steps. To further minimize interference from hydrogen ions, it is desirable to use solutions with pH values higher than 5. Improved mechanical and electrical properties can be achieved using more complex glasses containing various additives.

5.2.2 Liquid Membrane Electrodes

Liquid-membrane-type ISEs, based on water-immiscible liquid substances impregnated in a polymeric membrane, are widely used for direct potentiometric measurements (18–20). Such ISEs are particularly important because they permit direct measurements of several polyvalent cations as well as certain anions. The polymeric membrane [commonly made of plasticized poly(vinyl chloride) (PVC)] separates the test solution from the inner compartment, containing a standard solution of the target ion (into which a silver–silver chloride wire is dipped). The filling solution usually contains a chloride salt of the primary ion, as desired for establishing the potential of the internal silver–silver chloride wire electrode. The membrane-active (recognition) component can be an ion exchanger or a neutral macrocyclic compound. The selective extraction of the target ion at the sample–membrane interface creates the electrochemical phase boundary potential. The membranes are commonly prepared by dissolving the recognition element, a plasticizer (e.g., *o*-nitrophenyl ether, which provides the properties of liquid phase), and the PVC in a solvent such as tetrahydrofuran. (The recognition element is usually present in 1–3% amount.) Slow (overnight) evaporation of the solvent leaves a flexible membrane of 50–200 μm thickness, which can be cut (with a cork borer) and mounted on the end of plastic tube. The ion-discriminating ability (and hence the selectivity coefficient) depends not only on the nature of the recognition element but also on the exact membrane composition, including the membrane solvent and the nature and content of the plasticizer. The extraction properties of the membrane can be further improved by adding ion-pairing agents to the plasticizer. The PVC matrix provides mechanical strength and permits diffusion of analytes to the recognition sites. The hydrophobic nature of the membrane prevents leaching of the sensing element and the plasticizer into the aqueous sample solution, and thus extends the operational lifetime. Different methacrylic–acrylic copolymers were suggested as alternative to PVC (21). Such polymers require no plasticizer and facilitate the covalent attachment of crown-ether recognition elements.

It was shown that leaching of the primary ion (from the internal electrolyte solution) leads to its higher activity at the layer adjacent to the membrane (relative to the bulk sample), and hence to increased detection limits of carrier-based liquid membrane electrodes (22–26). These fluxes maintain a micromolar activity in the proximity of the membrane–solution interface, even if the sample contains virtually no primary ions. Such a localized accumulation of ions makes it impossible to measure dilute samples and restricts the detection limits to the micromolar range. In addition, this leak limits the selectivity coefficients to $\sim 10^{-4}$. By choosing an internal electrolyte with low activity of the primary ion and preventing it from leaking it is possible to greatly lower the detection limits by up to six orders of magnitude down to the picomolar (ppt) range (23–26). Several schemes have thus been suggested for minimizing biases due to ion fluxes through the membrane and improving the detection limits. One way to accomplish this is to add a hydrophilic complexing agent (such as EDTA) or ion exchanger to the inner solution of the membrane. Such adjustment of the inner solution can be combined with membranes that are less sensitive to concentration gradients. Another way to avoid leaching of primary ions from the membrane into the sample is to apply a very small external current that generates a steady flux of cations toward the inner compartment of the ISE (27). This procedure is attractive because it is easier to change the required current (rather than adjusting the inner solution). Such current-induced galvanostatic elimination of undesired leaching is displayed in Figure 5.9. Inward fluxes (i.e., siphoning of the primary ion from the sample) may also result from significant ion exchange at the inner side of the membrane. This can lead to depletion of the primary ion from the sample side of the membrane and hence to a super-Nernstian response along with poor practical detection limits. Accordingly, fluxes in either direction should be avoided. Active research in various laboratories is currently elucidating these interfacial ion fluxes relative to the development of potentiometric sensors for trace analysis. Covalent bonding of the ionophore to a polymer backbone has also been shown useful for addressing the adverse effect of zero-current ion fluxes and for improving the detection limits (28). Such covalent attachment also extends the lifetime of the corresponding ISE.

5.2.2.1 Ion Exchanger Electrodes One of the most successful liquid membrane electrodes is selective toward calcium. Such an electrode relies on the ability of phosphate ions to form stable complexes with the calcium ion. It uses a liquid cation exchanger, consisting of an aliphatic diester of phosphoric acid

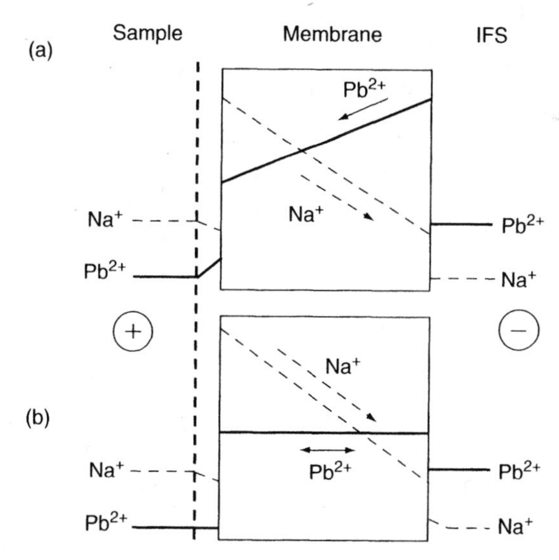

Figure 5.9 Elimination of primary ion leaching from the inner filling solution across the membrane, into the sample, by applying a negative current. Hypothetical concentration profiles in a lead-selective membrane before (a) and after (b) applied negative current. Arrows indicate the direction of the net ion fluxes. (Reproduced with permission from Ref. 25.)

[$(RO)_2PO_2^-$ with R groups in the C_8–C_{16} range], that possesses high affinity for calcium ions. The ion exchanger is held in a porous, plastic filter membrane that separates the test solution from the inner compartment, containing a standard calcium chloride solution (Fig. 5.10). The preferential uptake of calcium ions into the membrane can thus be represented as

$$Ca^{2+} + 2(RO)_2PO_2^- \rightleftharpoons [(RO)_2PO_2]_2Ca \tag{5.14}$$

The resulting cell potential is given by

$$E_{cell} = K + \frac{0.059}{2} \log a_{Ca} \tag{5.15}$$

Calcium activities as low as 5×10^{-7} M can be measured, with selectivity coefficients of $K_{Ca,Mg}$ and $K_{Ca,K}$ of 0.02 and 0.001, respectively. Such potential response is independent of the pH over the pH range 5.5–11.0. Above pH 11, $Ca(OH)^+$ is formed, while below 5.5, protons interfere. Because of its attractive response characteristics, the calcium ISE has proved to be a valuable tool for the determination of calcium ion activity in various biological fluids.

Anion exchangers, such as lipophilic quaternary ammonium salts (e.g., see Fig. 5.11) or phosphonium salts, have been employed for the preparation of

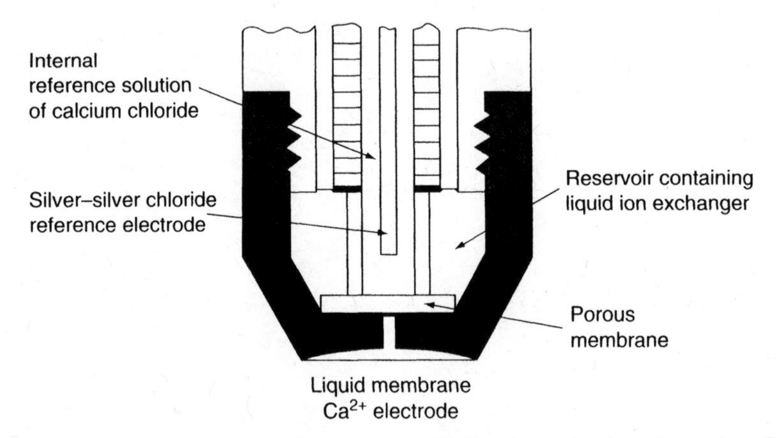

Figure 5.10 Schematic diagram of a calcium-ion-selective electrode.

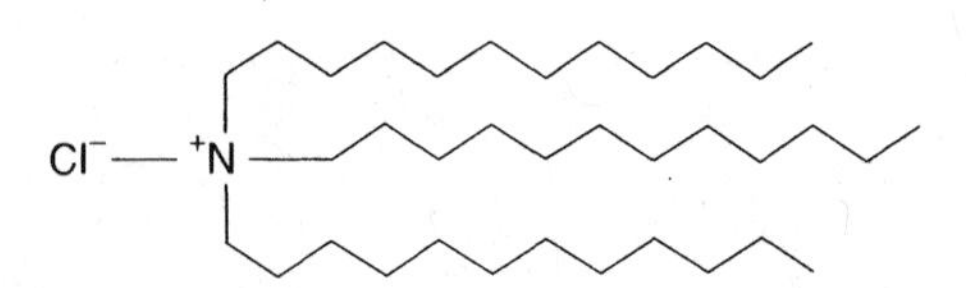

Figure 5.11 Quaternary alkyl ammonium chloride.

anion-selective sensors. The resulting ISEs usually lack an anion recognition function, and hence display anion selectivity corresponding to the anion partition into the supporting hydrophobic membrane. This gives rises to the following selectivity order, which is known as the *Hofmeister series*: large liphophilic anions > ClO_4^- > IO_4^- > SCN^- > I^- > NO_3^- > Br^- > Cl^- > HCO_3^- > $H_2PO_4^-$ (i.e., with maximum response to lipophilic anions) (29). Accordingly, several commercial sensors (e.g., NO_3^- "selective" electrodes), based on ion-exchange-type membranes, suffer an interference from liphohilic anions (e.g., ClO_4^-). Electrodes useful for nitrate (30), thiocyanate (31), and chloride (32) ions have thus been developed. Sensors responsive to anionic macromolecules have also been developed despite the greater difficulty in identifying appropriate membrane chemistry that yields a significant and selective response (33,34). A very successful example is the use of the quaternary ammonium salt tridodecylmethylammonium chloride (TDMAC) for detecting the clinically important drug heparin (33). Apparently, the polyionic heparin is favorably extracted into the membrane through ion-pairing interaction with the positively charged nitrogen atoms (Fig. 5.12). Such an extraction process results in a steady-state change in the phase boundary potential at the membrane–sample interface. Analogous potentiometric measurements of other macromolecular polyanionic (e.g., polyphosphates, DNA) or polycationic (e.g., protamine, polyarginine) species, based on the use of various lipophilic ion exchangers, have been reported (34,35). Ion exchange electrodes sensitive to large organic cations have also been described. For example, PVC membranes containing dinonylnaphthalenesulfonic acid (DNNS) have been used for the detection of drugs of abuse (e.g., opiate alkaloids) (36). Such organic-responsive elec-

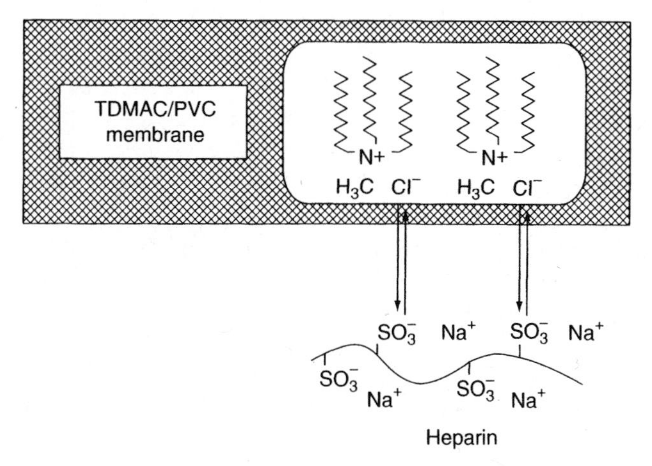

Figure 5.12 The recognition process occurring at the TDMAC/PVC membrane–sample interface used for measurements of heparin. (Reproduced with permission from Ref. 34.)

trodes, however, lack sufficient selectivity and are limited to simple samples, such as pharmaceutical formulations.

5.2.2.2 Neutral Carrier Electrodes In addition to charged liquid ion exchangers, liquid membrane electrodes often rely on the use of complex-forming neutral-charged carriers. Since the early 1980s much effort has been devoted to the isolation or synthesis of compounds containing cavities of molecule-sized dimensions. Such use of chemical recognition principles has made an enormous impact on widespread acceptance of ISEs. For example, most blood electrolyte determinations are currently being performed with ionophore-based sensors, either with centralized clinical analyzers or with decentralized disposable units.

Neutral carriers can be natural macrocyclic molecules or synthetic crown compounds (e.g, cyclic polyethers) capable of enveloping various target ions in their pocket. Electron-donor atoms, present in the polar host cavity, further facilitate and influence the interaction with the target ion. For example, while oxygen-containing crown ethers form stable complexes with alkali or alkali earth metals, sulfur-containing ones are best suited for binding heavy metals. The extent of this interaction is determined by the "best fit" mechanism, with larger ions that cannot fit in the molecular cavity, and smaller ones that are weakly coordinated. Often, a subunit group is added to the crown compound to impart higher selectivity (through a steric/blockage effect) and improved lipophilicity. Overall, these ionophores serve as reversible and reusable binding reagents that selectively extract the target analyte into the membrane. Such binding event creates the phase boundary potential at the membrane–sample interface. To ensure reversible binding, it is essential to keep the free energy of activation of the analyte–ionophore reaction sufficiently small (37). Molecular modeling techniques are being used to guide the design of ionophores toward target analytes. The specific design takes into consideration the selectivity demands imposed by clinical or environmental samples.

A host of carriers, with a wide variety of ion selectivities, have been proposed for this task. Most of them have been used for the recognition of alkali and alkaline metal cations (e.g., clinically relevant electrolytes). A classical example is the cyclic depsipeptide valinomycin (Fig. 5.13), used as the basis for the widely used ISE for potassium ion (38). This doughnut-shaped molecule has an electron-rich pocket in the center into which potassium ions are selectively extracted. For example, the electrode exhibits a selectivity for K^+ over Na^+ of approximately 30,000. The basis for the selectivity seems to be the fit between the size of the potassium ion (radius 1.33 Å) and the volume of the internal cavity of the macrocyclic molecule. The hydrophobic sidechains of valinomycin stretch into the lipophilic part of the membrane. In addition to its excellent selectivity, such an electrode is well behaved and has a wide working pH range. Strongly acidic media can be employed because the electrode is 18,000 times more responsive to K^+ than to H^+. A Nernstian response to potassium ion activities, with a slope of 59 mV/pK^+, is commonly observed

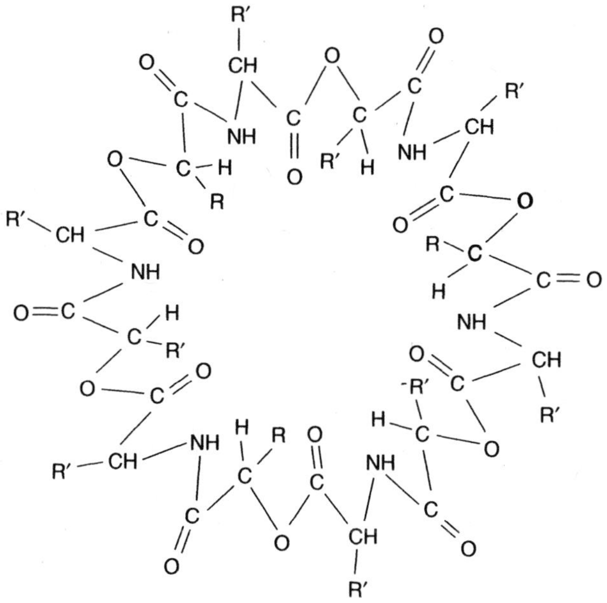

Figure 5.13 Valinomycin.

from 10^{-6} to 0.1 M. Such attractive performance characteristics have made the valinomycin ISE extremely popular for clinical analysis (with 200 million assays of blood potassium carried out annually in the United States using this device).

Many other cyclic and noncyclic organic carriers with remarkable ion selectivities have been used successfully as active hosts of various liquid membrane electrodes. These include the 14-crown-4-ether for lithium (39); 16-crown-5 derivatives for sodium; bis (benzo-18-crown-6 ether) for cesium; the ionophore ETH 1001[(*R*,*R*)-*N*,*N*′-bis(11-ethoxycarbonyl)undecyl-*N*,*N*′-4,5-tetramethyl-3,6-dioxaoctanediamide] for calcium; the natural macrocyclics nonaction and monensin for ammonia and sodium (40); respectively; the ionophore ETH 1117 for magnesium, calixarene derivatives for sodium (41) and lead (42); and macrocyclic thioethers for mercury and silver (43). Some common ionophores used for sensing different cations are displayed in Figure 5.14. The development of highly selective lithium electrodes for clinical monitoring of psychiatric patients (receiving lithium-based drugs) has been particularly challenging considering the large sodium interference. Similarly, highly selective ionophores for sodium are needed to address the large excess of potassium in the intracellular fluid. Neutral carrier ISEs have been successfully applied for environmental samples, including trace (subnanomolar) measurements of lead and lead speciation in various natural waters (42). For example, adjusting a lead-containing sample to various pH values has allowed reliable measurement of the fractions of uncomplexed lead (with a good correlation to the the-

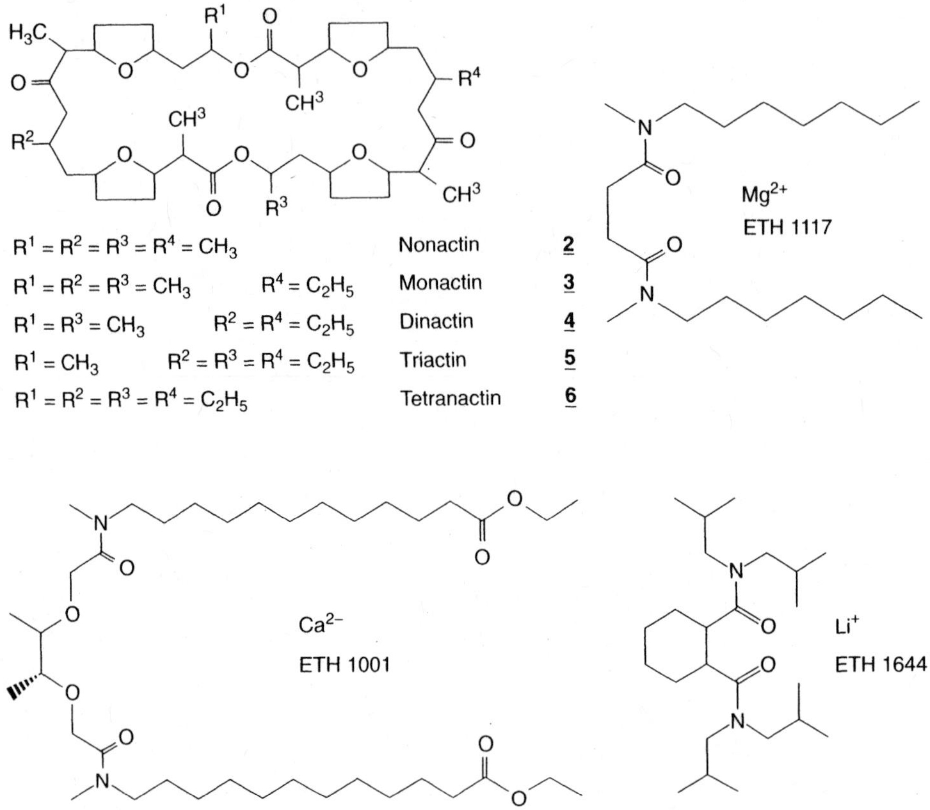

$R^1 = R^2 = R^3 = R^4 = CH_3$ Nonactin **2**

$R^1 = R^2 = R^3 = CH_3$ $R^4 = C_2H_5$ Monactin **3**

$R^1 = R^3 = CH_3$ $R^2 = R^4 = C_2H_5$ Dinactin **4**

$R^1 = CH_3$ $R^2 = R^3 = R^4 = C_2H_5$ Triactin **5**

$R^1 = R^2 = R^3 = R^4 = C_2H_5$ Tetranactin **6**

Mg^{2+} ETH 1117

Ca^{2-} ETH 1001

Li^+ ETH 1644

Figure 5.14 Structure of neutral carriers used in liquid membrane ion-selective electrodes.

oretical expectations; Fig. 5.15). The values obtained with various drinking-water samples were validated by inductively coupled plasma–mass spectrometry measurements.

Anion-selective liquid membrane electrodes have also been developed, based on the coordination of the anionic guest to host materials, such as metallophorphyrin or hydrophobic vitamin B_{12} derivatives, alkyltin compounds or macrocyclic polyamines (see Refs. 44–48 and Fig. 5.16). Such biomimetrically designed ionophores offer effective sensing of inorganic and organic anions, such as thiocyanate, carbonate, salicylate, phosphate, or adenosine nucleotides. Unlike anion exchanger electrodes, these anion sensors display selectivity patterns greatly different from the Hofmeister sequence (due to the direct interaction of the host with the specific anion). Often, this interaction involves an exchange of the coordinated anion at the metal center of the organometallic ionophore with the target anion in the sample solution. A review in 1998 describes in detail individual carrier-based ISEs, according to the analyte for

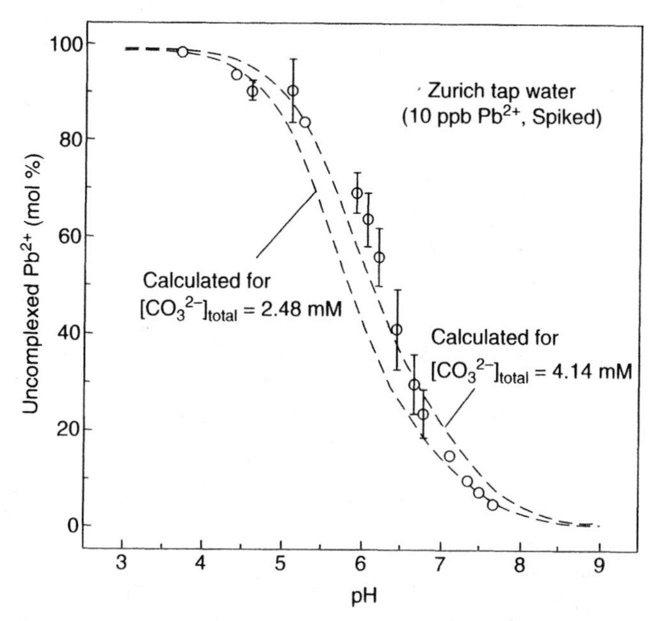

Figure 5.15 Direct potentiometic speciation of lead ion in Zurich tapwater as a function of pH. Dotted lines: expected from the known anions in the water sample and their complex equilibria with lead ion for two different carbonate concentrations. (Reproduced with permission from Ref. 42.)

which they have been developed (49). Many exciting developments based on novel host–guest chemistry (e.g., recognition by steric shapes) are anticipated in the near future.

5.2.3 Solid-State Electrodes

Considerable work has been devoted to the development of solid membranes that are selective primarily to anions. The solid-state membrane can be made of single crystals, polycrystalline pellets, or mixed crystals. The resulting solid-state membrane electrodes have found use in a great number of analytical applications.

An example of a very successful solid-state sensor is the fluoride-ion-selective electrode. Such a single-crystal device is by far the most successful anion-selective electrode. It consists of a LaF_3 crystal and an internal electrolyte solution (consisting of 0.1 M NaF and 0.1 M KCl, and containing the Ag/AgCl wire). The LaF_3 crystal is doped with some EuF_2 to provide vacancies ("holes") at anionic sites (Fig. 5.17). Such a solid-state membrane derives its selectivity from restriction of the movement of all ions, except the fluoride of interest. The latter moves by migration through the crystal lattice

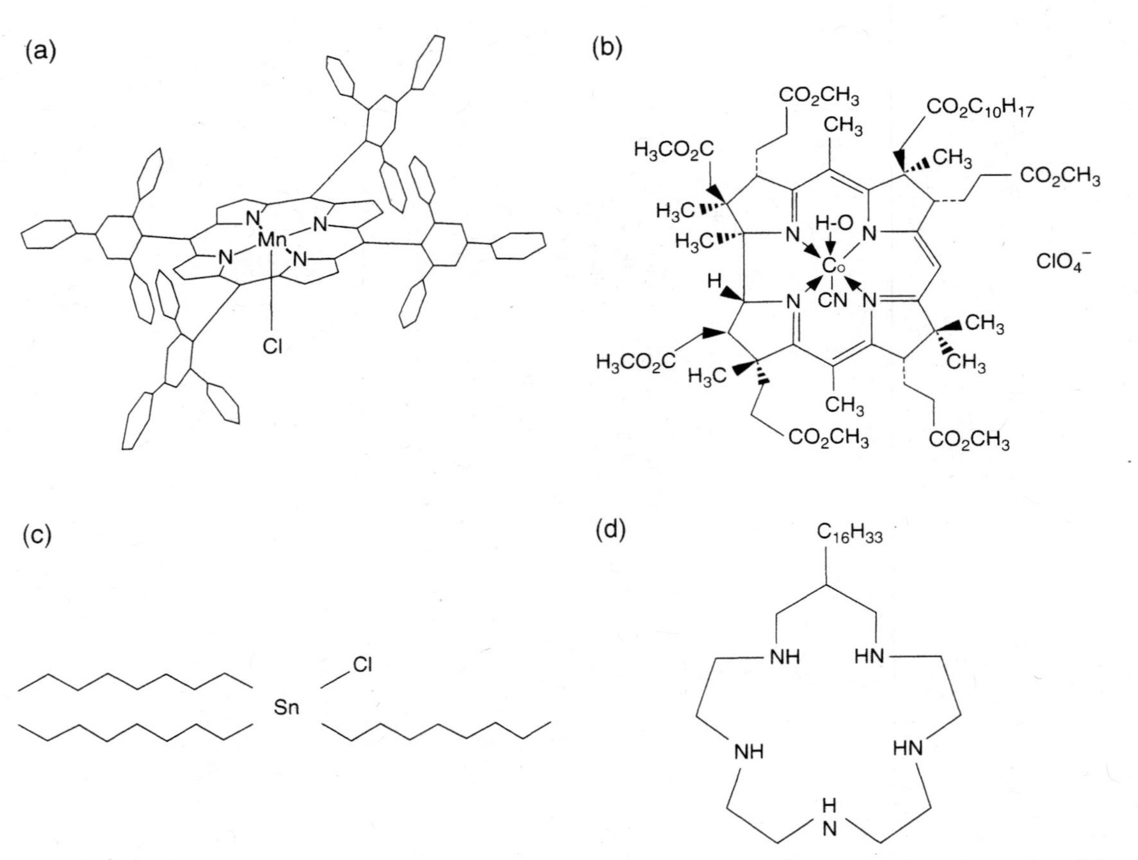

Figure 5.16 Structures of some chemical species useful for devising anion-selective electrodes: (a) Mn(III) porhyrin; (b) vitamin B_{12} derivative; (c) tri-*n*-octyltin chloride; (d) lipophilic polyamine macrocyclic compound.

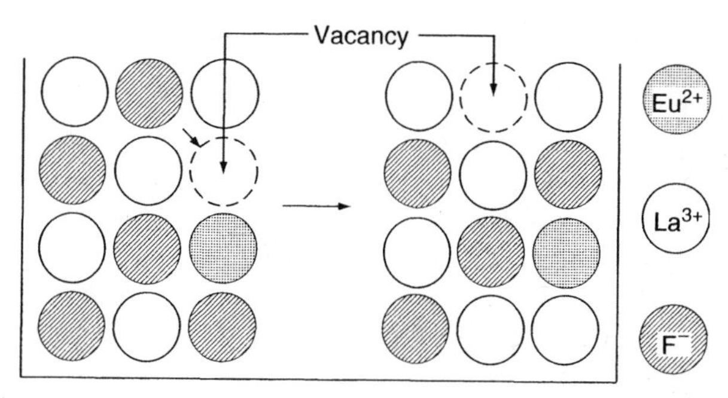

Figure 5.17 Migration of the fluoride ion through the LaF_3 lattice (doped with EuF_2). The vacancies created within the crystal cause jumping of neighboring F^- into the vacancy.

(by jumping from one vacancy defect to another), thus establishing the desired potential difference. A Nernstian response

$$E = K - 0.059 \log a_{F^-} \tag{5.16}$$

is obtained down to ~10^{-6} M. The only interfering ion (due to similarity in size and charge) is OH^-, for which the selectivity coefficient (K_{F^-/OH^-}) is 0.1. Hence, the electrode is limited to use over the pH range of 0–8.5. The electrode exhibits at least a 1000:1 preference for fluoride over chloride or bromide ions.

Other useful solid-state electrodes are based on silver compounds (particularly silver sulfide). Silver sulfide is an ionic conductor, in which silver ions are the mobile ions. Mixed pellets containing Ag_2S–AgX (where X = Cl, Br, I, SCN) have been successfully used for the determination of one of these particular anions. The behavior of these electrodes is basically determined by the solubility products involved. The relative solubility products of various ions with Ag^+ thus dictate the selectivity [i.e., $k_{ij} = K_{SP(Agi)}/K_{SP(Agj)}$]. Consequently, the iodide electrode (membrane of Ag_2S/AgI) displays high selectivity over Br^- and Cl^-. In contrast, the chloride electrode suffers from severe interference from Br^- and I^-. Similarly, mixtures of silver sulfide with CdS, CuS, or PbS provide membranes that are responsive to Cd^{2+}, Cu^{2+}, or Pb^{2+}, respectively. A limitation of these mixed-salt electrodes is that the solubility of the second salt must be much greater than that of silver sulfide. A silver sulfide membrane by itself responds to either S^{-2} or Ag^+ ions, down to the 10^{-8} M level.

Sensors for various halide ions can also be prepared by suspending the corresponding silver halide in an inert support material, such as silicone rubber (50). Such support material provides a flexible, heterogeneous membrane with resistance to cracking and swelling. The resulting membrane is called a heterogeneous or precipitate-impregnated membrane. For example, a chloride-selective electrode is based on a heterogeneous membrane prepared by polymerizing monomeric silicone rubber in the presence of an equal weight of silver chloride particles. A 0.5-mm-thick disk of this heterogeneous membrane is sealed to the bottom of a glass tube; potassium chloride and a silver wire are then placed in the tube. The sensitivity of such an electrode is limited by the solubility of silver chloride. Chloride concentrations from 5×10^{-5} to 1.0 M can be measured. Such an electrode operates over the pH range 2–12, and at temperatures between 5 and 5000°C. Ion-selective electrodes for thiocyanate (SCN^-) or cyanide (CN^-) can be prepared in a similar fashion. Such electrodes rely on a "corrosion" reaction between the silver halide (AgX) and the target ion, for example

$$AgX + 2CN^- \rightarrow Ag(CN)_2^- + X^- \tag{5.17}$$

(Safety considerations dictate that cyanide measurements be carried out in strongly basic media.) The interference mechanism with silver-based solid-state ISEs differs from that of ISEs described earlier. Depending on the K_{SP} value, an excess of the interfering ion may result in its deposit as silver salt on the membrane surface. Removal of the interfering film (by scrubbing) is thus required for restoring the electrode activity. Table 5.1 lists some solid-state electrodes from a commercial source, along with their dynamic range and major interferences.

5.2.4 Coated-Wire Electrodes and Solid-State Electrodes Without an Internal Filling Solution

Coated-wire electrodes (CWEs), introduced by Freiser in the mid-1970s, are prepared by coating an appropriate polymeric film directly onto a conductor (Fig. 5.18). The ion-responsive membrane is commonly based on poly(vinyl

TABLE 5.1 Characteristics of Solid-State Crystalline Electrodes

Analyte Ion	Concentration Range (M)	Major Interferences
Br^-	10^0–5×10^{-6}	CN^-, I^-, S^{2-}
Cd^{2+}	10^{-1}–1×10^{-7}	Fe^{2+}, Pb^{2+}, Hg^{2+}, Ag^+
Cl^-	10^0–5×10^{-5}	CN^-, I^-, Br^-, S^{2-}
Cu^{2+}	10^{-1}–1×10^{-8}	Hg^{2+}, Ag^+, Cd^{2+}
CN^-	10^{-2}–1×10^{-6}	S^{2-}
F^-	Saturated to 1×10^{-6}	OH^-
I^-	10^0–5×10^{-8}	
Pb^{2+}	10^{-1}–1×10^{-6}	Hg^{2+}, Ag^+, Cu^{2+}
Ag^+/S^{2-}	Ag^+: 10^0–1×10^{-7} S^{2-}: 10^0–1×10^{-7}	Hg^{2+}
SCN	10^0–1×10^{-6}	I^-, Br^-, CN^-, S^{2-}

Source: From *Orion Guide to Ion Analysis*, Orion Research, Cambridge, MA. 1983. With permission.

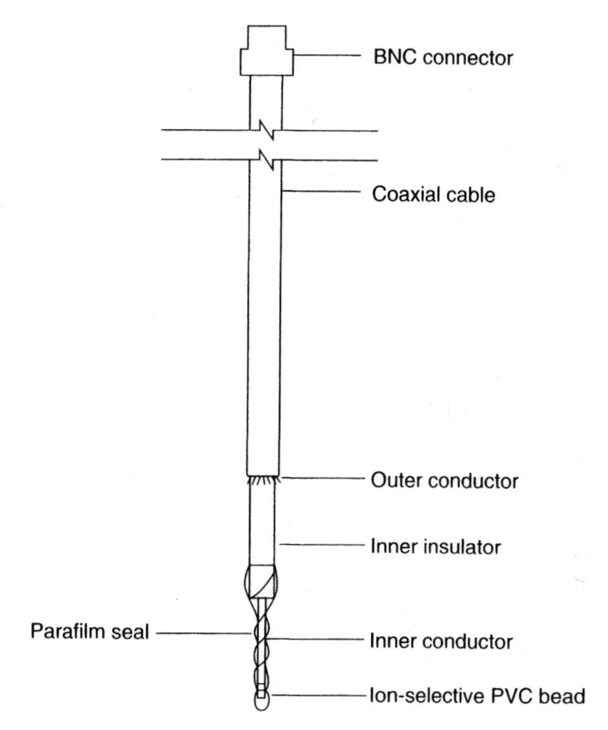

Figure 5.18 Coated-wire ion-selective electrode. (Reproduced with permission from Ref. 51.)

chloride), while the conductor can be metallic (Pt, Ag, Cu) or graphite-based of any conventional shape, such as wire or disk. The conductor is usually dipped in a solution of PVC and the active substance, and the resulting film is allowed to air-dry. Other polymers and modified polymers, including poly(acrylic acid) and modified poly(vinylbenzyl chloride), can also be useful for various applications. In addition to the miniaturization capability, CWEs are extremely simple, inexpensive, and easy to prepare and function well over the 10^{-5}–0.1 M concentration range. The exact mechanism of the CWE behavior continues to be a mystery, in view of the lack of internal reference components. Coated-wire electrodes may suffer from reproducibility and long-term stability (drifting potential) problems, resulting from the poorly defined contact and mechanism of charge transfer between the membrane coating and the conducting transducer. Nevertheless, such devices have been found useful for various important applications, provided that the electrodes are calibrated periodically. The determination of basic drugs, such as cocaine, methodane (52), amino acids (53), potassium, and sodium (54), represents some of the useful applications of CWE. The principles and applications of CWEs have been reviewed (4). New concepts for preparing CWEs appear to improve their analytical performance, particularly with respect to stability and reproducibility (through the achievement of thermodynamically defined interfaces). Such ability to eliminate the internal filling solution is currently receiving considerable interest in connection to mass production of potentiometric sensors and sensor arrays (see Section 6.3.2). Such ability offers great promise for eliminating steady-state ion fluxes (that lead to higher activity at the layer adjacent to the membrane) and hence to lower the detection limits compared to traditional ISE (25). Microfabricated (planar) ISE can also be designed with thin hydrogel layers, replacing the large-volume inner filling solution and addressing the stability and reproducibility limitations of CWEs. Such solid-state planar electrodes hold great promise for developing disposable ion sensors for decentralized applications ranging from home blood testing to on-site environmental monitoring.

Another route to address the potential stability of CWE and for mass-producing miniaturized ISE is to use an intermediate conducting polymer membrane between the conducting surface and the ion-selective membrane (55–57). This route gives solid-state ISE where the selectivity is determined by the ion-selective membrane, while the conducting polymer acts as the ion-to-electron transducer. Conducting polymers such as polypyrrole, polythiophene, or polyaniline have thus been shown useful for replacing the inner solution and preparing solid-state ISE. Such conducting-polymer-based sensors demonstrate high stability similar to that of conventional ISE (with an internal filling solution) (56). Doping the conducting polymer layer with an appropriate complexing agent can be used to lower the detection limits down to the nanomolar range (57). In certain cases, it is possible to incorporate the ion recognition sites directly into the conducting polymer matrix and hence eliminate the external ion-selective membrane (55).

5.3 ON-LINE, ON-SITE, AND IN VIVO POTENTIOMETRIC MEASUREMENTS

Various on-line monitoring systems can benefit from the inherent specificity, wide scope, dynamic behavior, and simplicity of ISEs. In particular, ISEs have been widely used as detectors in high-speed automated flow analyzers, such as air-segmented or flow injection systems (58,59). For example, Figure 5.19 shows the flow injection determination of physiologically important potassium in serum, using a tubular potassium selective electrode, at a rate of 100 samples per hour. Even higher throughputs, reaching 360 samples per hour, have been employed in connection with air-segmented flow systems (61). Such analyzers are now being routinely employed in most hospitals for the high-speed determination of physiologically important cationic electrolytes (e.g., K^+, Na^+, Ca^{2+}, Mg^{2+}, and H^+) or anions (e.g., Cl^-) in body fluids. The corresponding ISEs are usually placed in series, along a zigzag-shaped flow channel. Additional advantages accrue from the coupling of arrays of potentiometric detectors with chemometric (statistical) procedures (see Section 6.4).

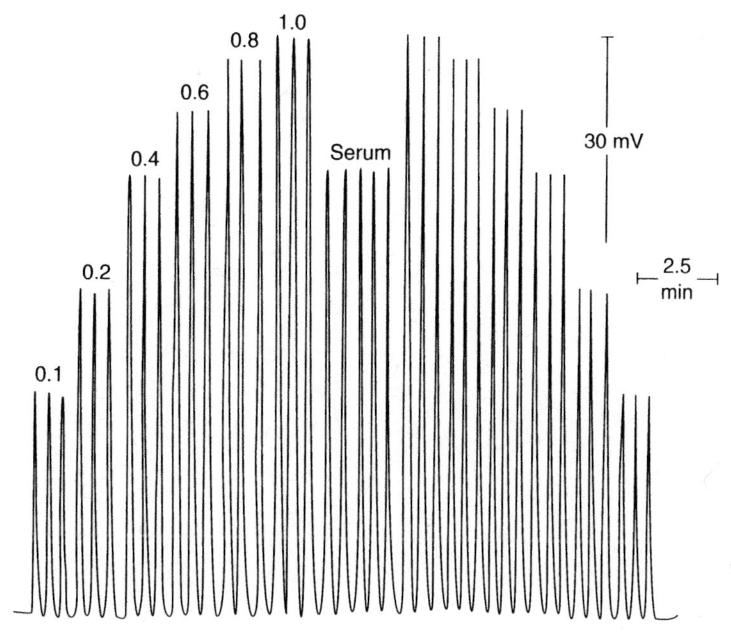

Figure 5.19 Flow injection potentiometric determination of potassium in serum. (Reproduced with permission from Ref. 60.)

The transient nature of flow injection potentiometric measurements (e.g., see Fig. 5.19) nicely addresses the potential drift problem common to analogous batch measurements. Such peak profiles are very reproducible, with any point on the peak easily related to the analyte activity. It can also be exploited for enhancing the selectivity by operating under kinetic (rather than equilibrium) control. Such kinetic selectivity reflects the faster rate of exchange of the primary ion (compared to interferents). Several designs of low-volume potentiometric flow detectors have been reported (60–64). The simplest design consists of an ISE fitted tightly with a plastic cap, with an inlet and outlet for the flowing stream (Fig. 5.20). The reference electrode is usually placed downstream from the ISE. It can also be immersed in a parallel (potassium chloride) flowing stream. Other common detector designs include the flow-through tubular ISE (used in Fig. 5.19), and tangential or wall-jet ISEs. Multi-ion detectors, based on ion-sensitive field effect transistors (discussed in Section 6.3) have been combined with miniaturized micromachined flow injection systems (63). Such coupling offers improved response times and reduced consumption of samples and reagents. Miniaturized arrays of multiple polymer membrane and solid-state ISE (for K^+, Na^{2+}, Ca^{2+}, Mg^{2+}, NH_4^+, Ba^{2+}, NO_3^-, Cl^-, and Li^+) have been developed for measuring terrestrial soil samples obtained in NASA missions to Mars (65).

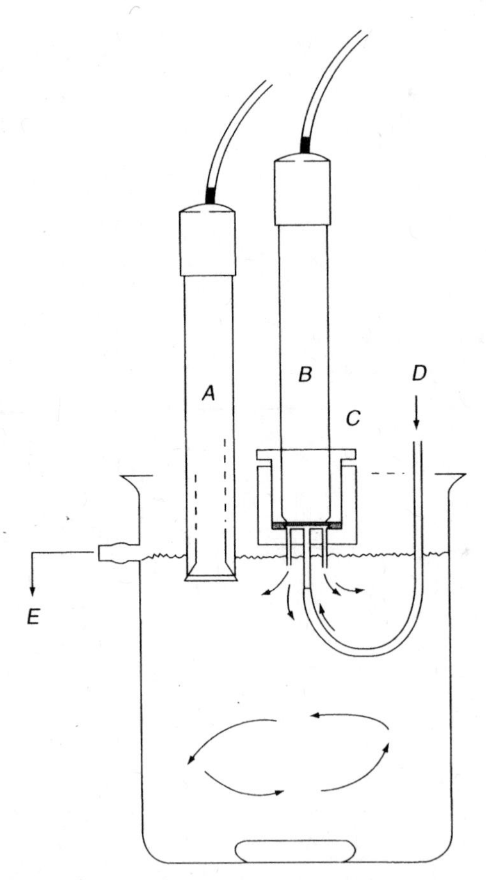

Figure 5.20 Flow-through potentiometric cell cap design: *A*, reference electrode; *B*, iodide electrode; *C*, flow-through cap; *D*, inlet; *E*, outlet. (Reproduced with permission from Ref. 62.)

In addition to automated analysis, ISEs can be used to detect ionic species in chromatographic effluents. Particularly powerful is the coupling of modern ion chromatography with potentiometric detection (66). Similarly, liquid membrane microelectrodes have been used as a small dead-volume detector in open tubular column liquid chromatography (67). Miniaturization has also permitted the adaptation of ISEs as on-column detectors for capillary-zone electrophoresis in connection with femtoliter detection volumes (68,69). The small dimensions in capillary electrophoresis require proper attention to the positioning of the ISE detector. Both micropipette and coated-wire ISEs have been useful for this task, with the latter offering a simplified electrode alignment (69). Micropipette ISEs have also been used as tips in scanning electrochemical microscopy (see Ref. 70; also Section 2.3, below).

Potentiometric microelectrodes are very suitable for in vivo real-time clinical monitoring of blood electrolytes, intracellular studies, in situ environmental surveillance, or industrial process control. For example, Simon's group described the utility of a system for on-line measurements of blood potassium ion concentration during an open-heart surgery (71); Buck and coworkers (72) reported on the use of flexible planar electrode arrays for the simultaneous in vivo monitoring of the pH and potassium ion in the porcine beating heart during acute ischemia (Fig. 5.21). Miniaturized catheter-type ISE sensors, such

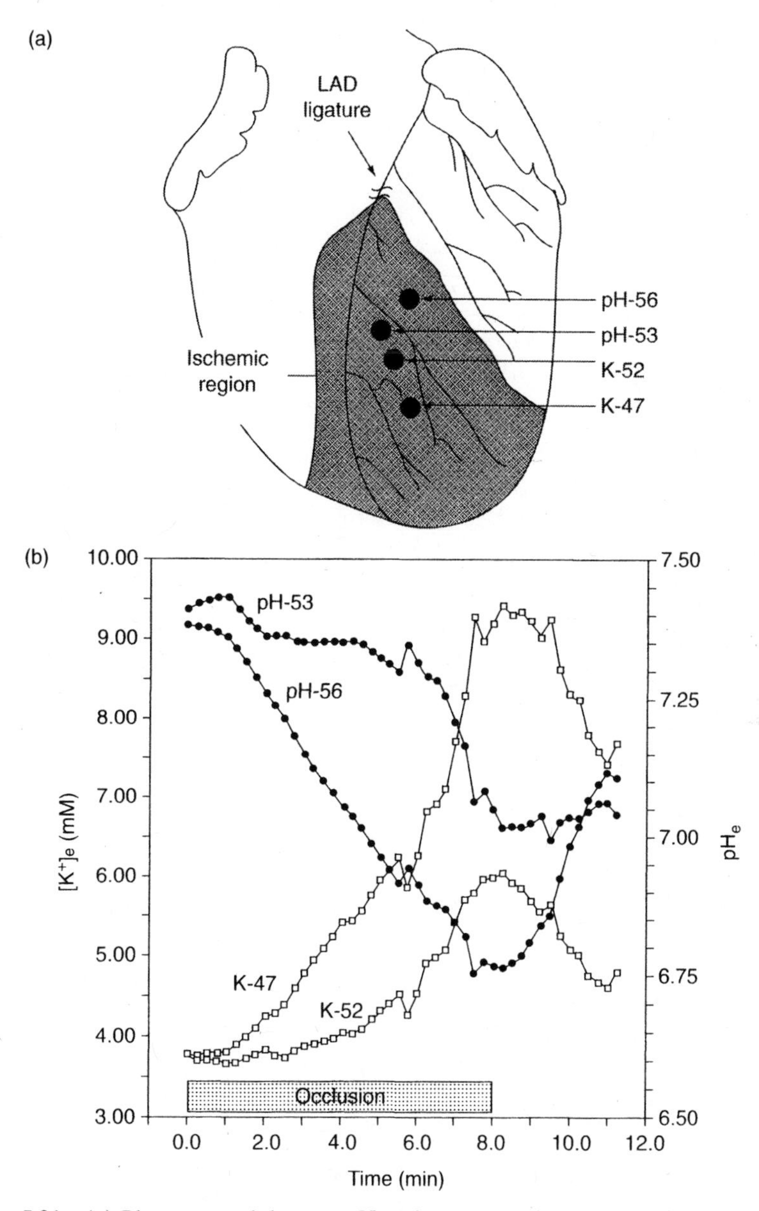

Figure 5.21 (a) Placement of the two pH and two potassium sensors (0.5 mm diameter) in the porcine heart; (b) recorded fall in the pH and increased potassium activity. (Reproduced with permission from Ref. 72.)

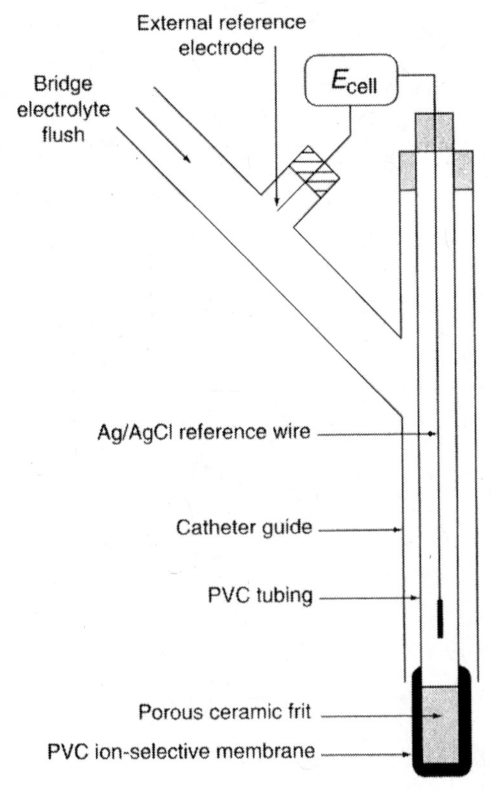

Figure 5.22 Miniaturized ISE catheter sensor for continuous monitoring of blood electrolytes. (Reproduced with permission from Ref. 73.)

as the implantable probe shown in Figure 5.22, represent the preferred approach for routine clinical in vivo monitoring of blood electrolytes. For these intravascular measurements the reference electrode is placed outside the artery (in the external arm of the catheter), thus obviating biocompatability and drift problems associated with its direct contact with the blood. Diamond's group developed an array of miniaturized chloride, sodium, and potassium ISEs for point-of-care analysis of sweat in connection to non-invasive diagnosis of cystic fibrosis (74). Note that shrinking ISE has a minimal effect on ISE behavior, since ISE response is size-independent. Yet, further miniaturization of ISEs to the nanometer domain is limited by the electrical resistance of the bulk liquid membrane.

EXAMPLES

Example 5.1 Calculate the relative error (in proton concentration) that would occur if the pH of a 1×10^{-2} M NaOH solution were measured with a pH glass electrode ($k_{H,Na} = 10^{-10}$, assuming an activity coefficient of 1.0).

Solution The concentration of the interfering sodium ion is 1×10^{-2} M, while that of the target proton is

$$[H^+] = K_w/[OH^-] = 10^{-14}/10^{-2} = 1 \times 10^{-12}\ M$$

From Eq. (5.6), we thus obtain

$$E_{cell} = K + 0.059 \log[1 \times 10^{-12} + 10^{-10} \times (1 \times 10^{-2})]$$

The relative error in concentration is thus

$$[10^{-10} \times (1 \times 10^{-2})/10^{-12}] \times 100 = 100\%$$

Example 5.2 The following potentials were observed for a calcium electrode immersed in standard calcium solutions:

$[Ca^{2+}]$ (M)	E (mV)
1×10^{-5}	100
1×10^{-4}	129
1×10^{-3}	158

What potential is expected for a calcium concentration of 5×10^{-4} M? (Assume an activity coefficient of 1.0.)

Solution Plotting E against $\log[Ca^{2+}]$ gives a straight line with a slope of 29 mV/decade and an intercept of 245 mV. A calcium concentration of 5×10^{-4} M thus yields

$$E = K + 29 \log[5 \times 10^{-4}] = 245 - 95.7 = 149.3\ mV$$

Example 5.3 Calculate the error in millivolts that would occur if a solution containing 5×10^{-5} M F^- (pH 10) were measured with a fluoride ISE ($k_{F,OH}$ = 0.1).

Solution The concentration of the interfering hydroxyl ion at pH = 10 can be obtained as follows:

$$[OH^-] = K_w/[H^+] = 10^{-14}/10^{-10} = 10^{-4}\ M$$

From Eq. (5.4) we obtain

$$E_{cell} = K - 0.059 \log(5 \times 10^{-5} + 0.1 \times 10^{-4}) = K + 0.249$$

In the absence of hydroxyl ion, we would obtain

$$E_{\text{cell}} = K - 0.059\log(5\times10^{-5}) = K + 0.254$$

Therefore the error (in mV) would be

$$\text{Error} = K + 249 - (K + 254) = -5\,\text{mV}$$

Example 5.4 A student calibrated a Mg^{2+} ion-selective electrode using two standard solutions at 25°C and constant ionic strength and obtained the following results:

$[Mg^{2+}]$ (M)	E (mV)
1×10^{-3}	142
1×10^{-4}	113

What is the concentration of the test solution that gave a potential reading of 125 mV under the same conditions?

Solution Plotting E versus $\log[Mg^{2+}]$ gives a straight line; the magnesium concentration (2.6×10^{-4} M) corresponding to the 125-mV reading can be read directly from the axis.

Example 5.5 Calculate the error caused by sodium ion, $a_{\text{Na}} = 0.01$, in the measurement of lithium, $a_{\text{Li}} = 0.001$, using a lithium-ion-selective electrode ($k_{\text{Li,Na}} = 0.06$).

Solution From Eq. (5.6), we thus obtain

$$E = K + 0.059\log[0.001 + 0.06(0.01)] = K - 0.165\,\text{V}$$

Without sodium, the potential is

$$E = K + 0.059\log(0.001) = K - 0.177\,\text{V}$$

The error is $(0.012/0.177) \times 100 = 6.8\%$.

PROBLEMS

5.1 Discuss the structural requirements for designing selective ionophores for ISE work. Give examples of such structures.

5.2 Explain (using one or multiple equations) why a highly selective ISE is not always sufficient for accurate potentiometric measurements.

5.3 Explain (using one or multiple equations) why the sodium ISE is more sensitive than the calcium one.

5.4 Describe the source of error in pH measurements using the glass pH electrode.

5.5 Describe the response mechanism of the fluoride-ion-selective electrode. Explain why the OH^- is the major interfering ion in F^- ISE measurements.

5.6 Give an example of a successful ISE for measuring a macromolecular polyanionic compound.

5.7 A major advance in ionophore-based ISEs is the finding that their detection limits can be lowered from the micromolar range to the nanomolar or picomolar level. Discuss recent developments and the new understanding that led to such dramatic improvements in the detectability.

5.8 Discuss the significance of the selectivity coefficient of an ISE. How would you determine its value?

5.9 Give an example of a successful ISE for measuring a macromolecular polyanionic compound.

5.10 Use the Nikolski–Eisenman equation to explain why lowering the detection limit requires careful attention to the selectivity of the resulting ISE.

5.11 Explain how the presence of magnesium ion can influence the response of a calcium ISE.

5.12 By how many millivolts will the potential of a calcium ISE change on transferring the electrode from a 1×10^{-3} M $CaCl_2$ solution to a 1×10^{-2} M $CaCl_2$ one?

5.13 Explain why small uncertainties in the measured cell potential can cause large error in the response of ISEs.

5.14 Discuss the major sources of error in potentiometric measurements.

REFERENCES

1. Koryta, J., *Ion, Electrodes and Membranes*, Wiley, New York, 1982.
2. Morf, W. E., *The Principles of Ion-Selective Electrodes and of Membrane Transport*, Elsevier, Amsterdam, 1981.
3. Covington, A. K., ed., *Ion-Selective Electrode Methodology*, CRC Press, Boca Raton, FL, 1979.
4. Freiser, H., ed., *Ion Selective Electrodes in Analytical Chemistry*, Plenum Press, New York, Vol. 1, 1978; Vol. 2, 1980.

5. Lakshminarayanaiah, N., *Membrane Electrodes*, Academic Press, New York, 1976.
6. Buck, P. P., *CRC Crit. Rev. Anal. Chem.* **5**, 323 (1976).
7. Ammann, D.; Morf, W.; Anker, P.; Meier, P.; Pret, E.; Simon, W., *Ion Select. Electrode. Rev.* **5**, 3 (1983).
8. Convington, A. K., *CRC Crit. Rev. Anal. Chem.* **3**, 355 (1974).
9. Nikolskii, B. P., *Acta Physiochim. USSR* **7**, 597 (1937).
10. Umezawa, Y., ed., *Handbook of Ion-Selective Electrodes: Selectivity Coefficients*, CRC Press, Boca Raton, FL, 1990.
11. Bakker, E., *Electroanalysis* **9**, 7 (1997).
12. Buck, R. P.; Lindner, E., *Pure Appl. Chem.* **66**, 2527 (1995).
13. Lewenstam, A.; Maj-Zurawska, M.; Hulanicki, A., *Electroanalysis* **3**, 727 (1991).
14. Bakker, E.; Meyerhoff, M. E., *Anal. Chim. Acta* **416**, 121 (2000).
15. Cremer, M., *Z. Biol.* (Munich) **47**, 562 (1906).
16. Eisenman, G., ed., *Glass Electrodes for Hydrogen and Other Cations*, Marcel Dekker, New York, 1976.
17. Bates, R. G., *Determination of pH: Theory and Practice*, Wiley, New York, 1973.
18. Anker, P.; Ammann, D.; Simon, W., *Mikrochim. Acta* **I**, 237 (1983).
19. Oesch, U.; Ammann, D.; Simon, W., *Clin. Chem.* **38**, 1448 (1986).
20. Johnson, R. D.; Bachas, L. G., *Anal. Bioanal. Chem.* **376**, 328 (2003).
21. Heng, L. Y.; Hall, E. A. H., *Anal. Chem.* **72**, 42 (2000).
22. Mathison, S.; Bakker, E., *Anal. Chem.* **70**, 303 (1998).
23. Sokalski, T.; Ceresa, A.; Zwicki, T.; Pretsch, E., *J. Am. Chem. Soc.* **119**, 11347 (1997).
24. Bakker, E.; Buhlmann, P.; Pretsch, E., *Electroanalysis* **11**, 915 (1999).
25. Bakker, E.; Pretsch, E., *Anal. Chem.* **74**, 420A (2002).
26. Pergel, E.; Gyurcsanyi, R.; Toth, K.; Lindner, E., *Anal. Chem.* **73**, 4249 (2001).
27. Lindner, E.; Gyurcsanyi, R.; Buck, R. P., *Anal. Chem.* **72**, 1127 (2000).
28. Puntener, M.; Vigassy, T.; Baier, E.; Ceresa, A.; Pretsch, E., *Anal. Chim. Acta* **503**, 187 (2004).
29. Holfmeister, F., *Arch. Exp. Pathol. Pharmakol.* **24**, 247 (1888).
30. Hulanicki, A.; Maj-Zurawska, M.; Lewandowski, R., *Anal. Chim. Acta* **98**, 151 (1978).
31. Coetzee, C. J.; Freiser, H., *Anal. Chem.* **40**, 207 (1968).
32. Walker, J. L., *Anal. Chem.* **43**, 89A (1971).
33. Ma, S.; Yang, V.; Meyerhoff, M., *Anal. Chem.* **64**, 694 (1992).
34. Fu, B.; Bakker, E.; Yum, J.; Wang, E.; Yang, V.; Meyerhoff, M. E., *Electroanalysis* **7**, 823 (1995).
35. Ye, Q.; Meyerhoff, M. E., *Anal. Chem.* **73**, 332 (2001).
36. Martin, C. R.; Freiser, H., *Anal. Chem.* **52**, 562 (1980).
37. Pretsch, E.; Badertscher, M.; Welti, M.; Maruizumi, T.; Morf, W.; Simon, W., *Pure Appl. Chem.* **60**, 567 (1988).
38. Morf, W. E.; Ammann, D.; Simon, W., *Chimica* **28**, 65 (1974).
39. Gadzekpo, V.; Hungerford, J.; Kadry, A.; Ibrahim, Y.; Christian, G., *Anal. Chem.* **57**, 493 (1985).

40. Kedem, O.; Loebel, E.; Furmansky, M., Gen. Offenbach (Patent) 2027128 (1970).
41. Cadogan, A.; Diamond, D.; Smyth, M.; Deasy, M.; McKervey, A.; Harris, S., *Analyst* **114**, 1551 (1989).
42. Ceresa, A.; Bakker, E.; Hattendorf, B.; Gunther, D.; Pretsch, E., *Anal. Chem.* **73**, 343 (2001).
43. Brzozka, Z.; Cobben, P.; Reinhaudt, D.; Edema, J.; Buter, J.; Kellogg, R., *Anal. Chim. Acta* **273**, 1139 (1993).
44. Wuthier, U.; Viet Pham, H.; Zund, R.; Welti, D.; Funck, R. J.; Bezegh, A.; Ammann, D.; Pretsch, E.; Simon, W., *Anal. Chem.* **56**, 535 (1984).
45. Tohda, K.; Tange, M.; Odashima, K.; Umezawa, Y.; Furuta, H.; Sessler, J., *Anal. Chem.* **64**, 960 (1992).
46. Park, S.; Matuszewski, W.; Meyerhoff, M.; Liu, Y.; Kadish, K., *Electroanalysis* **3**, 909 (1991).
47. Glazier, S.; Arnold, M., *Anal. Chem.* **63**, 754 (1991).
48. Chang, Q.; Park, S. B.; Kliza, D.; Cha, G. S.; Yim, H.; Meyerhoff, M. E., *Am. Lab.* **10** Nov. (1990).
49. Buhlmann, P.; Pretsch, E.; Bakker, E., *Chem. Rev.* **98**, 1593 (1998).
50. Pungor, E.; Toth, K., *Pure Appl. Chem.* **31**, 521 (1972).
51. Martin, C. R.; Freiser, H., *J. Chem. Educ.* **57**, 152 (1980).
52. Cunningham, L.; Freiser, H., *Anal. Chim. Acta* **139**, 97 (1982).
53. James, H.; Carmack, G.; Freiser, H., *Anal. Chem.* **44**, 856 (1972).
54. Tamura, H.; Kimura, K.; Shono, T., *Anal. Chem.* **54**, 1224 (1982).
55. Bobacka, J.; Ivaska, A.; Lewenstam, A., *Electroanalysis* **15**, 366 (2003).
56. Bobacka, J., *Anal. Chem.* **71**, 4932 (1999).
57. Konopka, A.; Sokalski, T.; Michalska, A.; Lewenstam, A.; Maj-Turawska, M., *Anal. Chem.* **76**, 6410 (2004).
58. Toth, K.; Fucsko, J.; Lindner, E.; Feher, Z.; Pungor, E., *Anal. Chim. Acta* **179**, 359 (1986).
59. Trojanowich, M.; Matuszewski, W., *Anal. Chim. Acta* **138**, 71 (1982).
60. Meyerhoff, M. E.; Kovach, P. M., *J. Chem. Educ.* **60**, 766 (1983).
61. Alexander, P. S.; Seegopaul, P., *Anal. Chem.* **52**, 2403 (1980).
62. Llenado, R.; Rechnitz, G. A., *Anal. Chem.* **45**, 2165 (1973).
63. Van der Schoot, B.; Jeanneret, S.; van den Berg, A.; de Rooji, N., *Anal. Meth. Instrum.* **1**, 38 (1993).
64. Gyurcsanyi, R. E.; Rangisetty, N.; Clifton, S.; Pendley, B.; Lindner, E., *Talanta* **63**, 89 (2004).
65. Lukow, S. R.; Kounaves, S. P., *Electroanalysis* **17**, 1441 (2005).
66. Isildak, I.; Covington, A., *Electroanalysis* **5**, 815 (1993).
67. Manz, A.; Simon, W., *J. Chromatogr. Sci.* **21**, 326 (1983).
68. Nann, A.; Silverstri, I.; Simon, W., *Anal. Chem.* **65**, 1662 (1993).
69. Kappes, T.; Hauser, P. C., *Anal. Chem.* **70**, 2487 (1998).
70. Toth, K.; Nagy, G.; Wei, C.; Bard, A. J., *Electroanalysis* **7**, 801 (1995).

71. Osswald, H. F.; Asper, R.; Dimai, W.; Simon, W., *Clin. Chem.* **25**, 39 (1979).
72. Buck, R. P.; Cosorfet, V.; Lindner, E.; Ufer, S.; Madaras, M.; Johnson, T.; Ash, R.; Neuman, M., *Electroanalysis* **7**, 846 (1995).
73. Espadas-Torre, C.; Telting-Diaz, M.; Meyerhoff, M. E., *Interface* **41** (spring issue) (1995).
74. Lynch, A.; Diamond, D.; Leader, M., *Analyst* **125**, 2264 (2000).

6

ELECTROCHEMICAL SENSORS

A chemical sensor is a small device that can be used for direct measurement of the analyte in the sample matrix. Ideally, such a device is capable of responding continuously and reversibly and does not perturb the sample. By combining the sample handling and measurement steps, sensors eliminate the need for sample collection and preparation. Chemical sensors consist of a transduction element covered by a chemical or biological recognition layer. This layer interacts with the target analyte, and the chemical changes resulting from this interaction are translated by the transduction element into electrical signals.

The development of chemical sensors is currently (as of 2005) one of the most active areas of analytical research. Electrochemical sensors represent an important subclass of chemical sensors in which an electrode is used as the transduction element. Such devices hold a leading position among sensors presently available, have reached the commercial stage, and have found a vast range of important applications in the fields of clinical, industrial, environmental, and agricultural analyses. The field of sensors is interdisciplinary, and future advances are likely to occur from progress in several disciplines. Research into electrochemical sensors is proceeding in a number of directions, as described in the following sections. The first group of electrochemical sensors, the potentiometric ion-selective electrodes (based on "ionic receptors"), has been described in Chapter 5.

Analytical Electrochemistry, Third Edition, by Joseph Wang

6.1 ELECTROCHEMICAL BIOSENSORS

Electrochemical biosensors combine the analytical power of electrochemical techniques with the specificity of biological recognition processes. The aim is to biologically produce an electrical signal that relates to the concentration of an analyte. For this purpose, a biospecific reagent is either immobilized or retained at a suitable electrode, which converts the biological recognition event into a quantitative amperometric or potentiometric response. Such biocomponent–electrode combinations offer new powerful analytical tools that are applicable to many challenging problems. A level of sophistication and state-of-the art technology are commonly employed to produce easy-to-use, compact, and inexpensive devices. Advances in electrochemical biosensors are progressing in different directions. Two general categories of electrochemical biosensors may be distinguished, depending on the nature of the biological recognition process: biocatalytic devices (utilizing enzymes, cells, or tissues as immobilized biocomponents) and affinity sensors (based on antibodies, membrane receptors, or nucleic acids).

6.1.1 Enzyme-Based Electrodes

Enzymes are proteins that catalyze chemical reactions in living systems. Such catalysts are not only efficient but also extremely selective. Hence, enzymes combine the recognition and amplification steps, as needed, for many sensing applications.

Enzyme electrodes are based on the coupling of a layer of an enzyme with an appropriate electrode. Such electrodes combine the specificity of the enzyme for its substrate with the analytical power of electrochemical devices. As a result of such coupling, enzyme electrodes have been shown to be extremely useful for monitoring a wide variety of substrates of analytical importance in clinical, environmental, and food samples.

6.1.1.1 Practical and Theoretical Considerations The operation of an enzyme electrode is illustrated in Figure 6.1. The immobilized enzyme layer is chosen to catalyze a reaction, which generates or consumes a detectable species:

$$S + C \xrightarrow{\text{enzyme}} P + C' \tag{6.1}$$

where S and C are the substrate and coreactant (cofactor), and P and C′ are the corresponding products. The choice of the sensing electrode depends primarily on the enzymatic system employed. For example, amperometric probes are highly suitable when oxidase or dehydrogenase enzymes (generating electrooxidizable hydrogen peroxide or NADH species) are employed, pH–glass electrodes for enzymatic pathways which result in a change in pH, while gas

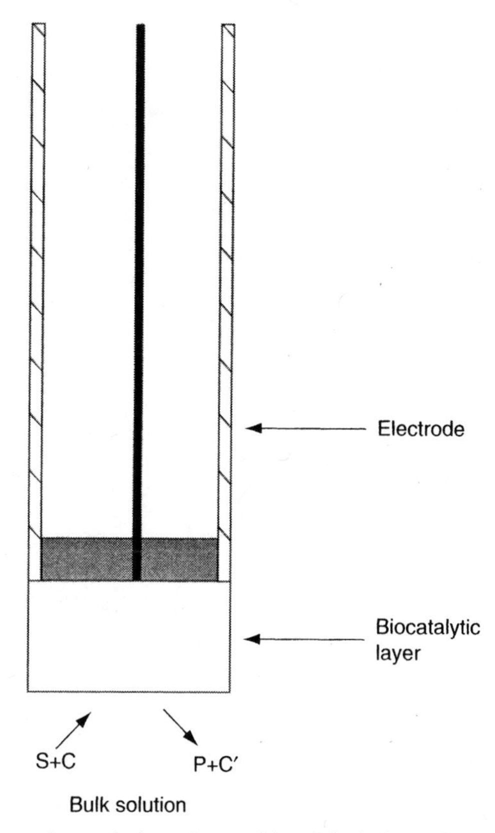

Figure 6.1 Enzyme electrode based on a biocatalytic layer immobilized on an electrode transducer.

(carbon dioxide) potentiometric devices will be the choice when decarboxylase enzymes are used.

The success of the enzyme electrode depends, in part, on the immobilization of the enzyme layer. The objective is to provide an intimate contact between the enzyme and the sensing surface while maintaining (and even improving) the enzyme stability. Several physical and chemical schemes can thus be used to immobilize the enzyme onto the electrode (Fig. 6.2). The simplest approach is to entrap a solution of the enzyme between the electrode and a dialysis membrane. Alternately, polymeric films (e.g., polypyrrole, Nafion) may be used to entrap the enzyme (via casting or electropolymerization). Additional improvements can be achieved by combining several membranes and/or coatings. Figure 6.3 displays a useful, and yet simple, immobilization based on trapping the enzyme between an inner cellulose acetate film and a collagen or polycarbonate membrane, cast at the tip of an amperometric transducer. Such coverage with a membrane/coating serves also to extend the linear range (via reduction of the local substrate concentration)

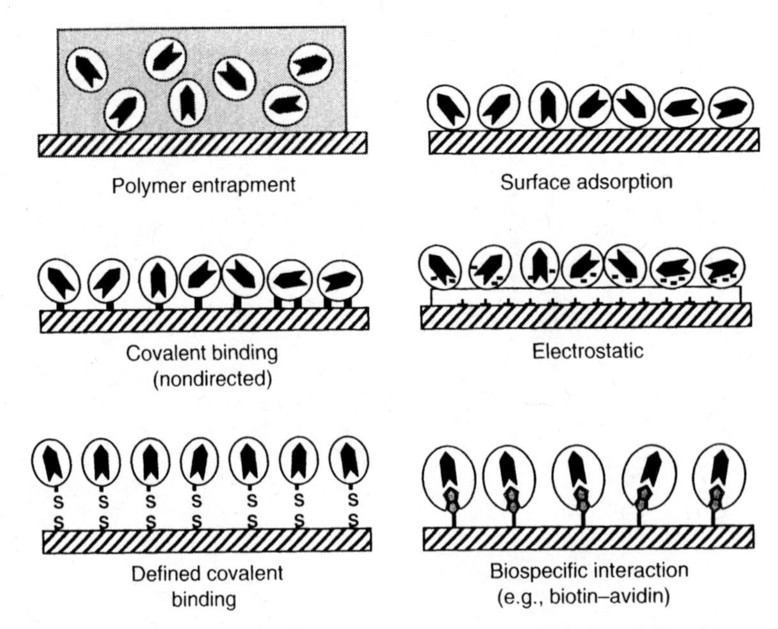

Figure 6.2 Methods for immobilizing enzymes onto electrode surfaces.

and to reject potential interferences (e.g., coexisting electroactive species or proteins). In chemical immobilization methods the enzyme is attached to the surface by means of a covalent coupling through a cross-linking agent (e.g., glutaraldehyde, amide). Covalent coupling may be combined with the use of functionalized thiolated monolayers for assembling multilayer enzyme networks on electrode surfaces (2). Biotin–avidin interactions can also be employed using streptavidin-coated surfaces and biotinylated enzymes (e.g., see Fig. 6.2). Other useful enzyme immobilization schemes include entrapment within a thick gel layer, low-temperature encapsulation onto sol-gel films, adsorption onto a graphite surface, incorporation (by mixing) within the bulk of three-dimensional carbon-paste or graphite–epoxy matrices (3,4), or electrochemical codeposition of the enzyme and catalytic metal particles (e.g., Pt, Rh). Such codeposition, as well as electropolymerization processes, are particularly suited for localizing the enzyme onto miniaturized sensor surfaces (5,6). The electropolymerization route can be accomplished by entrapping the enzyme within the growing film or anchoring it covalently to the monomer prior to the film deposition. Such an avenue can also reduce interferences and fouling of the resulting biosensors. The mixed-enzyme/carbon paste immobilization strategy is attractive for many routine applications, as it couples the advantages of versatility (controlled doping of several modifiers, e.g., enzyme, cofactor mediator), speed (due to close proximity of biocatalytic and sensing sites, and absence of membrane barriers), ease of fabrication, and renewability.

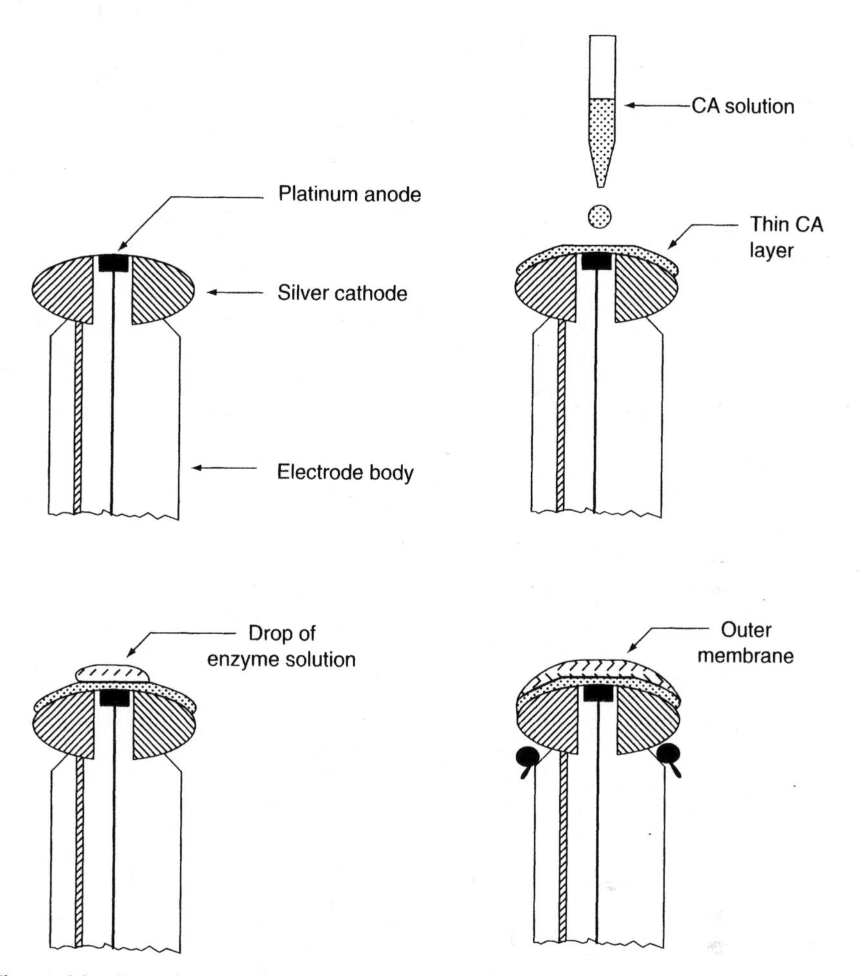

Figure 6.3 Steps in preparation of an amperometric enzyme electrode, with a simple enzyme immobilization by trapping between an inner cellulose acetate and outer collagen membrane, cast on the electrode body. (Reproduced with permission from Ref. 1.)

The immobilization procedure may alter the behavior of the enzyme (compared to its behavior in homogeneous solution). For example, the apparent parameters of an enzyme-catalyzed reaction (optimum temperature or pH, maximum velocity, etc.) may all be changed when an enzyme is immobilized. Improved stability may also accrue from the minimization of enzyme unfolding associated with the immobilization step. Overall, careful engineering of the enzyme microenvironment (on the surface) can be used to greatly enhance the sensor performance. More information on enzyme immobilization schemes can be found in several reviews (7, 8).

The response characteristics of enzyme electrodes depend on many variables, and an understanding of the theoretical basis of their function would help to improve their performance. Enzymatic reactions involving a single substrate can be formulated in a general way as

$$E + S \underset{k_{-1}}{\overset{k_1}{\rightleftharpoons}} ES \overset{k_2}{\rightarrow} E + P \tag{6.2}$$

In this mechanism, the substrate S combines with the enzyme E to form an intermediate complex ES, which subsequently breaks down into products P and liberates the enzyme. At a fixed enzyme concentration, the rate of the enzyme-catalyzed reaction V is given by the Michaelis-Menten equation:

$$V = V_m[S]/(K_m + [S]) \tag{6.3}$$

where K_m is the Michaelis-Menten constant and V_m is the maximum rate of the reaction. The term K_m corresponds to the substrate concentration for which the rate is equal to half of V_m. In the construction of enzyme electrodes, it is desirable to obtain the highest V_m and lowest K_m. Figure 6.4 shows the dependence of the reaction rate on the substrate concentration, together with the parameters K_m and V_m. The initial rate increases with substrate, until a nonlimiting excess of substrate is reached, after which additional substrate causes no further increase in the rate. Hence, a leveling off of calibration curves is expected at substrate concentrations above the K_m of the enzyme. Accordingly, low K_m values—while offering higher sensitivity—result in a narrower linear range (which reflects the saturation of the enzyme). The preceding discussion assumes that the reaction obeys the Michaelis-Menten kinetics theory. Experimentally, the linear range may exceed the concentration corresponding to K_m,

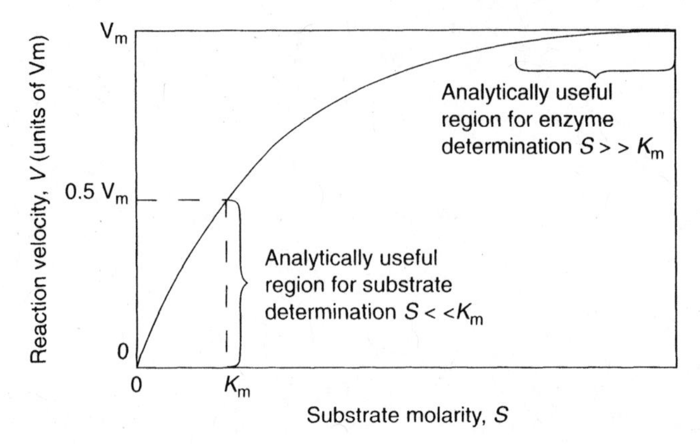

Figure 6.4 Dependence of the velocity of an enzyme-catalyzed reaction on the substrate concentration (at a constant level of the enzymatic activity).

because the local substrate concentration in the electrode containment region is often lower than the bulk concentration (as common with amperometric probes coated with diffusion-limiting membranes). The level of the cosubstrate may also influence the linear range.

Improved sensitivity and scope can be achieved by coupling two (or more) enzymatic reactions in a chain, cycling, or catalytic mechanism (9). For example, a considerable enhancement of the sensitivity of enzyme electrodes can be achieved by enzymatic recycling of the analyte in two-enzyme systems. Such an amplification scheme generates more than a stoichiometric amount of product and hence large analytical signals for low levels of the analyte. In addition, a second enzyme can be used to generate a detectable (electroactive) species, from a nonelectroactive product of the first reaction.

The most important challenge in amperometric enzyme electrodes is the establishment of satisfactory electrical communication between the active site of the enzyme and the electrode surface. Different mechanisms of electron transfer can be exploited for amperometric biosensing, including the use of natural secondary substrates, artificial redox mediators, or direct electron transfer (Fig. 6.5). The latter obviates the need for cosubstrates or mediators, holds promise for designing reagentless devices, and allows efficient transduction of the biorecognition event. Only a restricted number of enzymes have shown direct electron transfer reactions between the prosthetic group of the enzyme and electrodes (10). The challenges in establishing such direct electrical communication between redox enzymes and electrode surfaces have been reviewed (2,11,12).

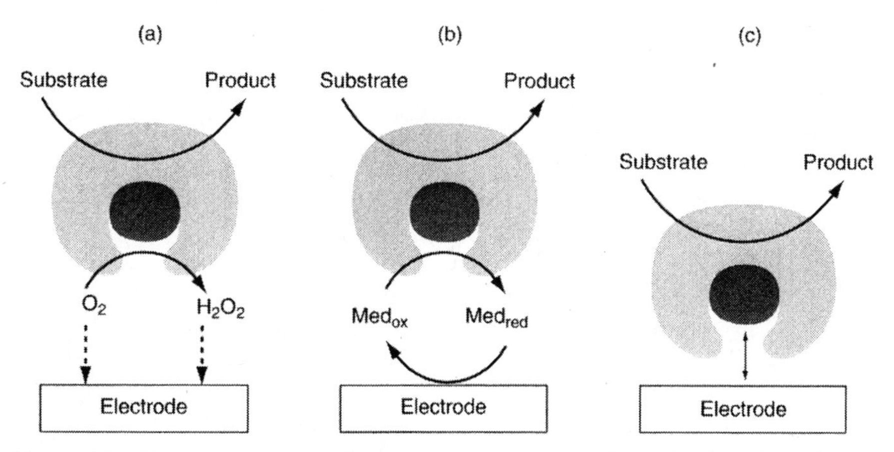

Figure 6.5 Three generations of amperometric enzyme electrodes based on the use of natural secondary substrate (a), artificial redox mediators (b), or direct electron transfer between the enzyme and the electrode (c).

6.1.1.2 *Enzyme Electrodes of Analytical Significance*

6.1.1.2.1 Glucose Sensors The determination of glucose in blood plays a crucial role in the diagnosis and therapy of diabetes. Electrochemical biosensors for glucose have played a key role in the move toward simplified wide-scale glucose testing, and have dominated the $5 billion/year diabetes monitoring market (13). The glucose amperometric sensor, developed by Updike and Hicks (14), represents the first reported use of an enzyme electrode. The electrode is commonly based on the entrapment of glucose oxidase (GOx) between polyurathene and permselective membranes on a platinum working electrode (Fig. 6.6). The liberation of hydrogen peroxide in the enzymatic reaction

$$\text{Glucose} + O_2 \xrightarrow{\text{glusoce oxidase}} \text{gluconic acid} + H_2O_2 \tag{6.4}$$

can be monitored amperometrically at the platinum surface:

$$H_2O_2 \xrightarrow{\text{electrode}} O_2 + 2H^+ + 2e^- \tag{6.5}$$

The multilayer membrane coverage (of Fig. 6.6) improves the relative surface availability of oxygen and excludes potential interferences (common at the potentials used for detecting the peroxide product). Electrocatalytic transducers based on Prussian Blue layers (15) or metallized carbons (16), which preferentially accelerate the oxidation of hydrogen peroxide, are also useful for minimizing potential interferences. The enzymatic reaction can also be followed by monitoring the consumption of the oxygen cofactor.

Further improvements can be achieved by replacing the oxygen with a nonphysiological (synthetic) electron acceptor, which is able to shuttle electrons from the flavin redox center of the enzyme to the surface of the working electrode. Glucose oxidase (and other oxidoreductase enzymes) do not directly transfer electrons to conventional electrodes because their redox centers are surrounded by a thick protein layer. Such insulating shell introduces a spatial separation of the electron donor–acceptor pair, and hence an intrinsic barrier to direct electron transfer, in accordance to the distance dependence of the electron transfer (ET) rate (17):

$$K_{et} = 10^{13} e^{-0.91(d-3)} e^{[-(\Delta G+\lambda)/4RT\lambda]} \tag{6.6}$$

where ΔG and λ correspond to the free and reorganization energies accompanying the electron transfer, respectively, and d is the actual electron transfer distance. The interfacial ET rate is thus dependent on the distance between the enzyme redox center and the electrode surface, that is, on the depth of the redox group inside the protein shell, and the orientation of the protein on the surface.

As a result of using artificial (diffusional) electron-carrying mediators, measurements become insensitive to oxygen fluctuations and can be carried

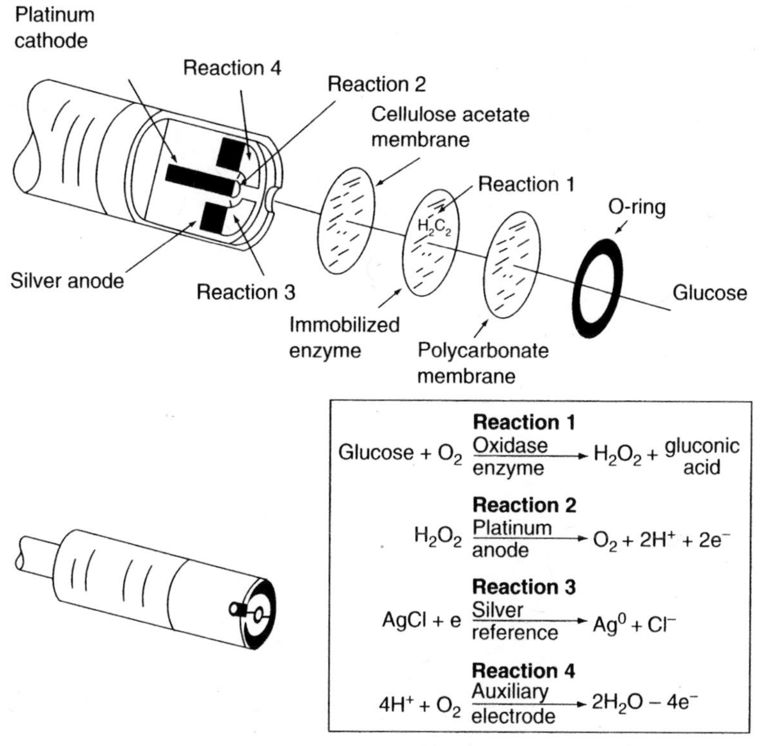

Figure 6.6 Schematic of a "first generation" glucose biosensor (based on a probe manufactured by YSI Inc.).

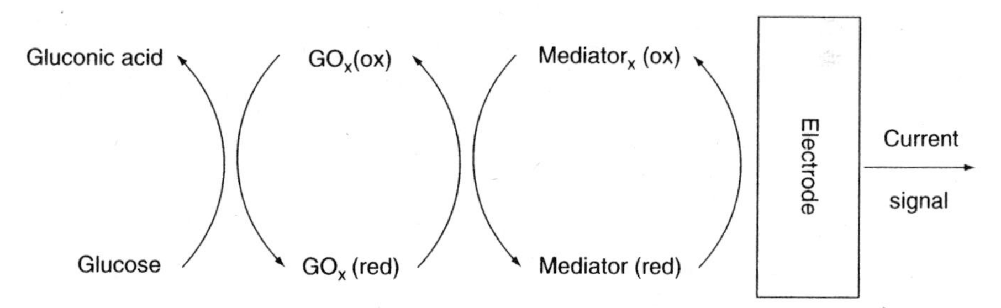

Figure 6.7 "Second generation" enzyme electrodes : sequence of events that occur in a mediated system (ox = oxidation; red = reduction). (Reproduced with permission from Ref. 19.)

out at lower potentials that do not provoke interfering reactions from coexisting electroactive species (Fig. 6.7). Many organic and organometallic redox compounds have been considered for this role of enzyme mediator (18–20). Some common examples are displayed in Figure 6.8. In particular, ferricyanide

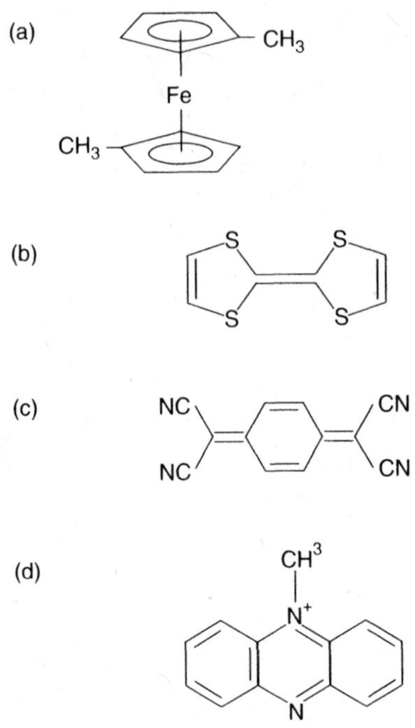

Figure 6.8 Chemical structures of some common redox mediators: (a) dimethyl ferrocene; (b) tetrathiafulvalene; (c) tetracyanoquinodimethane; (d) Meldola Blue.

and ferrocene derivatives (e.g., Fig. 6.8a) have been very successful for shuttling electrons from glucose oxidase to the electrode by the following scheme:

$$\text{Glucose} + \text{GOx}_{(\text{ox})} \rightarrow \text{gluconic acid} + \text{GOx}_{(\text{red})} \tag{6.7}$$

$$\text{GOx}_{(\text{red})} + 2\text{M}_{(\text{ox})} \rightarrow \text{GOx}_{(\text{ox})} + 2\text{M}_{(\text{red})} + 2\text{H}^{+} \tag{6.8}$$

$$2\text{M}_{(\text{red})} \rightarrow 2\text{M}_{(\text{ox})} + 2\text{e}^{-} \tag{6.9}$$

where $M_{(ox)}$ and $M_{(red)}$ are the oxidized and reduced forms of the mediator. This chemistry has led to the development of hand-held battery-operated meters for personal glucose monitoring in a single drop of blood (21). The single-use disposable strips used with these devices are usually made of polyvinyl chloride and a screen-printed carbon electrode containing a mixture of glucose oxidase and the mediator (Fig. 6.9). The screen-printing technology used for mass-scale production of this and similar biosensors, along with the ink-jet localization of the dry reagent layer, are discussed in Section 6.3. The control meter typically relies on a potential-step (chronoamperometric) operation. Other classes of promising mediators for glucose oxidase are quinone derivatives, ruthenium complexes, phenothiazine compounds, and organic conducting salts [particularly tetrathiafulvalene–tetracyanoquinodimethane

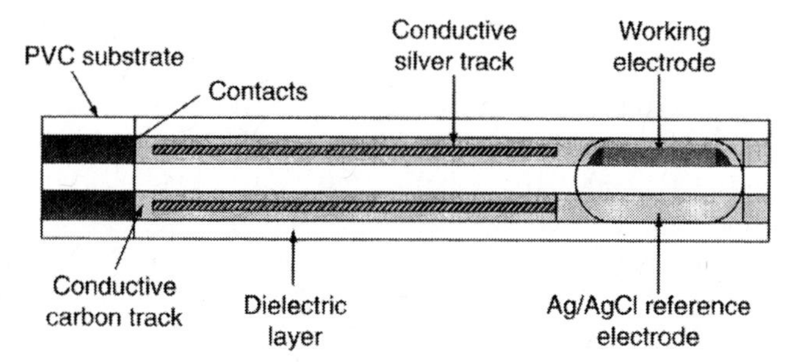

Figure 6.9 Schematic representation of a disposable glucose sensor strip. (Reproduced with permission from Ref. 20.)

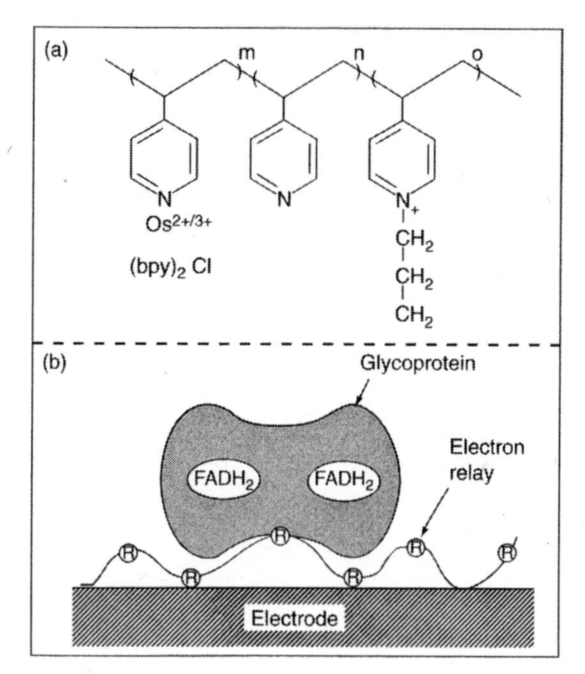

Figure 6.10 (a) Composition of an electron-relaying redox polymer and (b) use of the polymer for electrical "wiring" of an enzyme to the electrode surface. (Reproduced with permission from Ref. 22.)

(TTF-TCNQ)]. An elegant nondiffusional route for establishing electrical communication between GOx and the electrode is to "wire" the enzyme to the surface with a long polymer having a dense array of electron relays [e.g., osmium(bipyridyl) bound to poly(vinyl pyridine), Fig. 6.10a (22)]. Such a polymeric chain is flexible enough to fold along the enzyme structure (Fig. 6.10b). The resulting three-dimensional redox-polymer/enzyme network offers high current outputs and stabilizes the mediator to the surface. It has been successfully used in a commercial painless forearm blood glucose monitoring system. Nanoscale materials, such as gold nanoparticles or carbon nanotubes, have also shown to be extremely useful for "plugging" an electrode into GOx. (23,24). An even more elegant possibility is the chemical modification of the enzyme with the redox-active mediator (25). Glucose electrodes of extremely efficient electrical communication with the electrode can be generated by the enzyme reconstitution process (26). For this purpose, the flavin active center of GOx is removed to allow positioning of the electron-mediating ferrocene unit prior to reconstitution of the enzyme (Fig. 6.11). Ultimately, these and similar developments would lead to minimally invasive subcutaneously implanted (needle-type) and noninvasive devices for continuous real-time monitoring of glucose (27,28). Such probes would offer a tight control of diabetes, in connection with an alarm detecting hypo- or hyperglucemia or for a future closed-loop insulin release system (i.e., artificial pancreas). In addition to their biosensing utility, mediated enzyme electrodes (particularly those relying on electron-conducting redox polymers) have been shown extremely useful for increasing the power density of energy-producing biofuel cells (29,30). Such devices exploit the biocatalytic oxidation of biofuels, such as glucose, coupled to the enzymatic reduction of dissolved oxygen (by bilirubin oxidase or laccase), to generate electricity.

6.1.1.2.2 Ethanol Electrodes The reliable sensing of ethanol is of great significance in various disciplines. The enzymatic reaction of ethanol with the

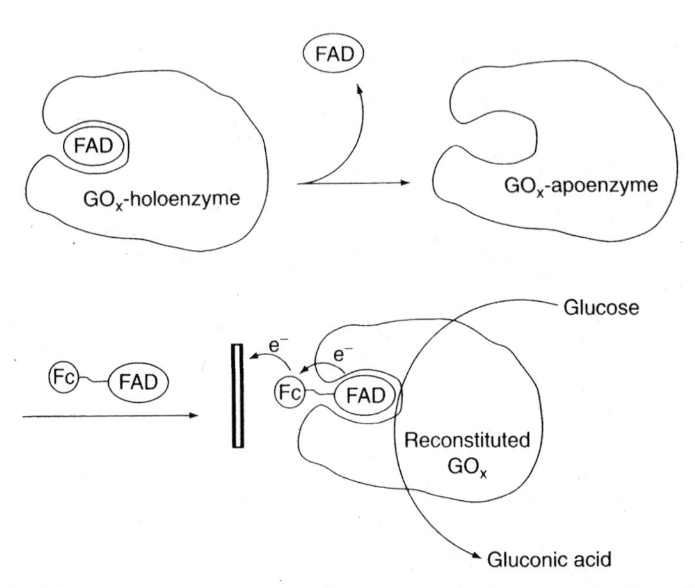

Figure 6.11 Electrical contacting of a flavoenzyme by its reconstitution with a relay FAD semi-synthetic cofactor. (Reproduced with permission from Ref. 2.)

cofactor nicotinamide adenine dinucleotide (NAD^+), in the presence of alcohol dehydrogenase (ADH)

$$C_2H_5OH + NAD^+ \xrightarrow{ADH} C_2H_5O + NADH \quad (6.10)$$

serves as a basis of amperometric sensors for ethanol (31). Reagentless devices based on the coimmobilization of ADH and NAD^+ to various carbon or platinum anodes are employed for this task (e.g., Fig. 6.12). NAD^+ is regenerated electrochemically by oxidation of the NADH, and the resulting anodic current is measured:

$$NADH \rightarrow NAD^+ + 2e^- + H^+ \quad (6.11)$$

To circumvent high overvoltage and fouling problems encountered with reaction (6.11) at conventional electrodes, much work has been devoted to the development of modified electrodes with catalytic properties for NADH. Immobilized redox mediators, such as the phenoxazine Meldola Blue or phenothiazine compounds, have been particularly useful for this task (32) (see also Fig. 4.17). Such mediation should be useful for many other dehydrogenase-based biosensors. High sensitivity and speed are indicated from the flow injection response illustrated in Figure 3.22. The challenges of NADH detection and the development of dehydrogenase biosensors have been reviewed (33). Alcohol biosensing can be accomplished also in the presence of alcohol oxidase, based on measurements of the liberated peroxide product.

6.1.1.2.3 Urea Electrodes The physiologically important substrate urea can be sensed on the basis of the following urease-catalyzed reaction:

$$NH_2CONH_2 + 2H_2O + H^+ \xrightarrow{urease} 2NH_4^+ + HCO_3^- \quad (6.12)$$

The electrode is an ammonium ion-selective electrode surrounded by a gel impregnated with the enzyme urease [Fig. 6.13 (34)]. The generated ammo-

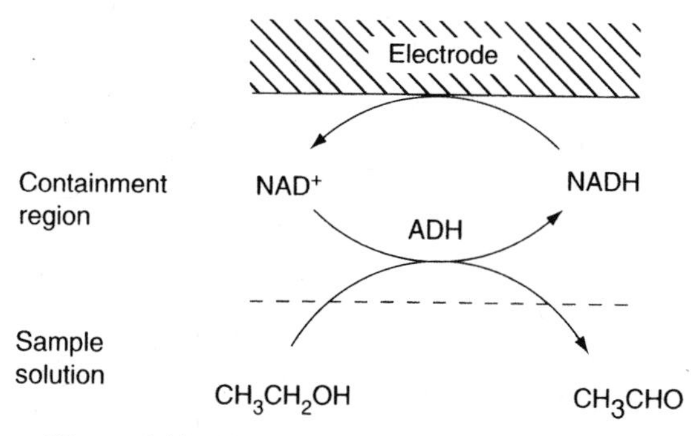

Figure 6.12 Reagentless ethanol bioelectrode.

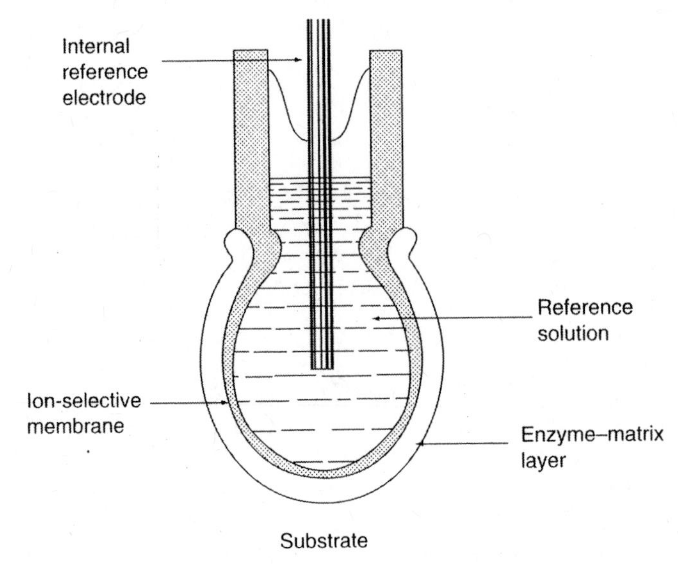

Figure 6.13 Urea electrode, based on the immobilization of urease onto an ammonium-ion-selective electrode.

TABLE 6.1 Some Common Enzyme Electrodes

Measured Species	Enzyme	Detected Species	Type of Sensing	Ref.
Cholesterol	Cholesterol oxidase	O_2	Amperometric	35
Creatinine	Creatinase	NH_3	Potentiometric gas sensing	36
			Amperometric	37
Lactate	Lactate dehydrogenase	NADH	Amperometric	38
	Lactate oxidase	H_2O_2	Amperometric	39
Penicillin	Penicillinase	H^+	Potentiometric	40
Phenol	Tyrosinase	Quinone	Amperometric	41
Salicylate	Salicylate hydroxylase	CO_2	Potentiometric gas sensing	42
Uric acid	Uricase	CO_2	Potentiometric gas sensing	43

nium ions are detected after 30–60 s to reach a steady-state potential. Alternately, the changes in the proton concentration can be probed with glass pH or other pH-sensitive electrodes. As expected for potentiometric probes, the potential is a linear function of the log[urea] in the sample solution.

Enzyme electrodes for other substrates of analytical significance have been developed. Representative examples are listed in Table 6.1. Further advances in enzyme technology, particularly the isolation of new and more stable enzymes, should enhance the development of new biocatalytic sensors. New opportunities (particularly assays of new environments or monitoring of hydrophobic analytes) accrued from the finding that enzymes can maintain their biocatalytic activity in organic solvents (44,45).

6.1.1.2.4 Toxin (Enzyme Inhibition) Biosensors Enzyme affectors (inhibitors and activators), which influence the rate of biocatalytic reactions, can also be measured. Sensing probes for organophosphate and carbamate pesticides, for the respiratory poisons cyanide or azide, or for toxic metals have thus been developed using enzymes such as acetylcholinesterase, horseradish peroxidase, or tyrosinase (46,47). The analytical information is commonly obtained from the decreased electrochemical response to the corresponding substrate (associated with the inhibitor–enzyme interaction). Pesticide measurements with cholinesterase systems often employ a bienzyme cholinesterase/choline oxidase system, in connection to amperometric monitoring of the liberated peroxide species. Changes in the substrate response can also be exploited for measuring the activity of enzymes.

6.1.1.3 Tissue and Bacteria Electrodes The limited stability of isolated enzymes, and the fact that some enzymes are expensive or even not available in the pure state, has prompted the use of cellular materials (plant tissues, bacterial cells, etc.) as a source of enzymatic activity (48). For example, the banana tissue (which is rich with polyphenol oxidase) can be incorporated by mixing within the carbon paste matrix to yield a fast-responding and sensitive dopamine sensor (Fig. 6.14). These biocatalytic electrodes function in a manner similar to that for conventional enzyme electrodes (i.e., enzymes present in the tissue or cell produce or consume a detectable species).

Other useful sensors rely on the coupling of microorganisms and electrochemical transducers. Changes in the respiration activity of the microorganism, induced by the target analyte, result in decreased surface concentration of electroactive metabolites (e.g., oxygen), which can be detected by the transducer.

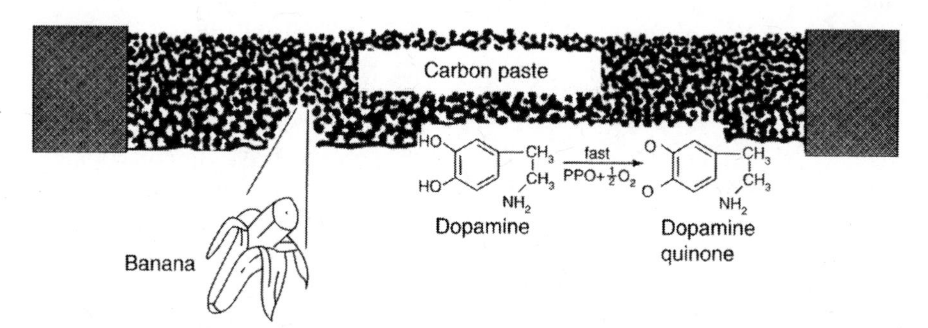

Figure 6.14 The mixed tissue (banana)–carbon paste sensor for dopamine. (Reproduced with permission from Ref. 49.)

6.1.2 Affinity Biosensors

Affinity electrochemical biosensors exploit selective binding of certain biomolecules (e.g., antibodies, receptors, or oligonucleotides) toward specific target species for triggering useful electrical signals. The biomolecular recognition process is governed primarily by the shape and size of the receptor pocket and the ligand of interest (the analyte). Such an associative process is governed by thermodynamic considerations (in contrast to the kinetic control exhibited by biocatalytic systems). The high specificity and affinity of biochemical binding reactions (such as DNA hybridization and antibody–antigen compexation) lead to highly selective and sensitive sensing devices. As will be shown in the following sections, electrochemical transducers are very suitable for detecting these molecular recognition events. Such devices rely on measuring the electrochemical signals resulting from the binding process.

6.1.2.1 Immunosensors Immunoassays are among the most specific of the analytical techniques, provide extremely low detection limits, and can be used for a wide range of substances. As research moves into the era of proteomic, such assays become extremely useful for identifying and quantitating proteins. Immunosensors are based on immunological reactions involving the shape recognition of the antigen (Ag) by the antibody (Ab) binding site to form the antibody/antigen (AbAg) complex:

$$Ab + Ag \rightleftharpoons AbAg \tag{6.13}$$

The antibody is a globular protein produced by an organism to bind to foreign molecules, namely, antigens, and mark them for elimination from the organism. The remarkable selectivity of antibodies is based on the stereospecificity of the binding site for the antigen, and is reflected by large binding constants (ranging from 10^5 to 10^9 L/mol). Antibody preparations may be monoclonal or polyclonal. The former are produced by a single clone of antibody-producing cells, and thus have the same affinity. Polyclonal antibodies, in contrast, are cheaper but possess varying affinities.

Electrochemical immunosensors, combining specific immunoreactions with an electrochemical transduction, have gained considerable attention (50–55). Such sensors are based on labeling of the antibody (or antigen) with an enzyme that acts on a substrate and generate an electroactive product that can be detected amperometrically. Enzyme immunosensors can employ competitive or sandwich modes of operation (Fig. 6.15). In competitive-type sensors, the sample antigen (analyte) competes with an enzyme-labeled antigen for antibody-binding sites on a membrane held on an amperometric or potentiometric sensing probe. After the reaction is complete, the sensor is washed to remove unreacted components. The probe is then placed in a solution containing the substrate for the enzyme, and the product or reactant of the biocatalytic reaction is measured. Because of the competitive nature of the

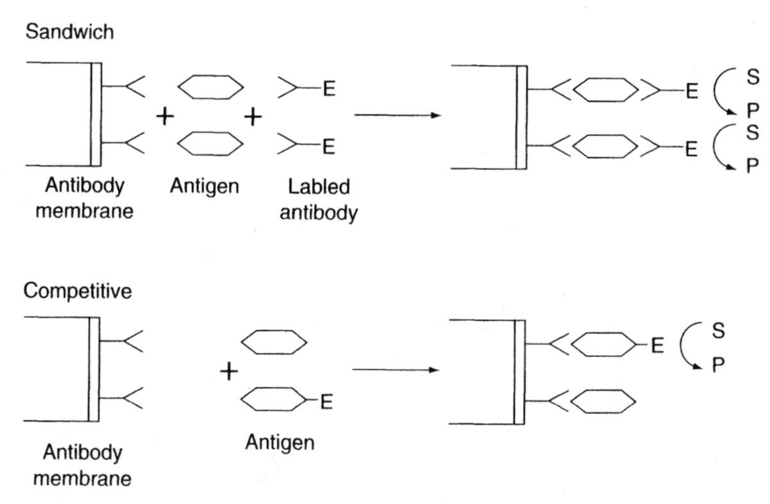

Figure 6.15 Enzyme immunosensors based on the competitive or sandwich modes of operation. (Reproduced with permission from Ref. 53.)

assay, the measurement signal is inversely proportional to the concentration of the analyte in the sample. Several enzymes, such as alkaline phosphatase, horseradish peroxidase, glucose oxidase, and catalase, have been particularly useful for this task.

Sandwich-type sensors are applicable for measuring large antigens that are capable of binding two antibodies. Such sensors utilize an antibody that binds the analyte antigen, which then binds the enzyme-labeled antibody. After removal of the nonspecifically adsorbed label, the probe is placed into the substrate-containing solution, and the extent of the enzymatic reaction is monitored electrochemically. Other types of immunosensors based on labeling the antigen or antibody with an electroactive tag (e.g., heavy metal or a ferrocene derivative), metal (gold) nanoparticle tracer, label-free capacitance, impedance or amperometric measurements, immobilizing antigen carrier conjugates at the tip of potentiometric electrodes, or amplifying the antigen–antibody complex equilibria by liposome lysis, are also being explored. For example, antibodies incorporated in conducting polymers have been shown to retain their affinity properties in connection with label-free pulsed amperometric measurements (56). Similarly, impedance spectroscopy (described in Section 2.5) offers a label-free electronic detection based on the increased interfacial electron transfer resistance associated with the formation of bioaffinity complexes (57). Changes in the conductivity of one-dimensional antibody-functionalized nanowires conjugated on binding of the target proteins can also lead to a powerful label-free electrical immunoassays (58).

Instead of immobilizing the antibody onto the transducer, it is possible to use a bare (amperometric or potentiometric) electrode for probing enzyme immunoassay reactions (59). In this case, the content of the immunoassay reaction vessel is injected into an appropriate flow system containing an electrochemical detector, or the electrode can be inserted into the reaction vessel. Remarkably low (femtomolar) detection limits have been reported in connection to the use of the alkaline phosphatase label (60,61). This enzyme catalyzes the hydrolysis of phosphate esters to liberate easily oxidizable phenolic products. Even lower detection limits can be achieved by coupling the electrochemical immunoassays with a dual-enzyme substrate regeneration (62). The use of gold nanoparticles has also been shown useful for highly sensitive immunoassays with stripping voltammetric detection of the dissolved gold (63).

More recent trends aim in the direction of fabricating electrochemical protein array systems (for detecting multiple protein targets) and miniaturization of such immunoassays. These include an electrochemical protein chip with an array of 36 platinum electrodes on a glass substrate (64) and electrical immunoassays using microcavity formats down to the zmol antigen level (65).

In addition to antibodies, it is possible to use artificial nucleic acids ligands, known as *aptamers*, for the selective detection of proteins. The tight binding properties make aptamers attractive candidates as molecular recognition elements in a wide range of bioassays and for the development of protein arrays. Electrochemistry has been shown useful for monitoring aptamer–protein interactions (66).

6.1.2.2 DNA Hybridization Biosensors

6.1.2.2.1 Background and Principles Nucleic acid recognition layers can be combined with electrochemical transducers to form new and important types of affinity biosensors. The use of nucleic acid recognition layers represents an exciting area in biosensor technology. Electrochemical DNA hybridization biosensors offer considerable promise for obtaining sequence-specific information in a simpler, faster, and cheaper manner, compared to traditional hybridization assays (67–71). Such strategies hold an enormous potential for clinical diagnosis of genetic or infectious diseases, for the detection of food-contaminating organisms, for early warning against biowarfare agents, for environmental monitoring, or in criminal investigations.

The basis for these devices is the DNA base pairing. Accordingly, these sensors rely on the immobilization of a relatively short [20–40-bp (basepair)] single-stranded DNA sequence (the "probe") on the transducer surface, which, on hybridization to a specific complementary region of the target DNA, gives rises to an electrical signal (Fig. 6.16). A wide range of chemistries have been exploited for monitoring electrochemically the DNA hybridization. These can be divided into two major principles, involving the use of labels generating an electrical signal or label-free protocols. The hybridization event can thus be detected via the increased current signal of an electroactive indicator (that preferentially binds to the DNA duplex), or due to captured enzyme or nanoparticle tags, or from other hybridization-induced changes in electro-

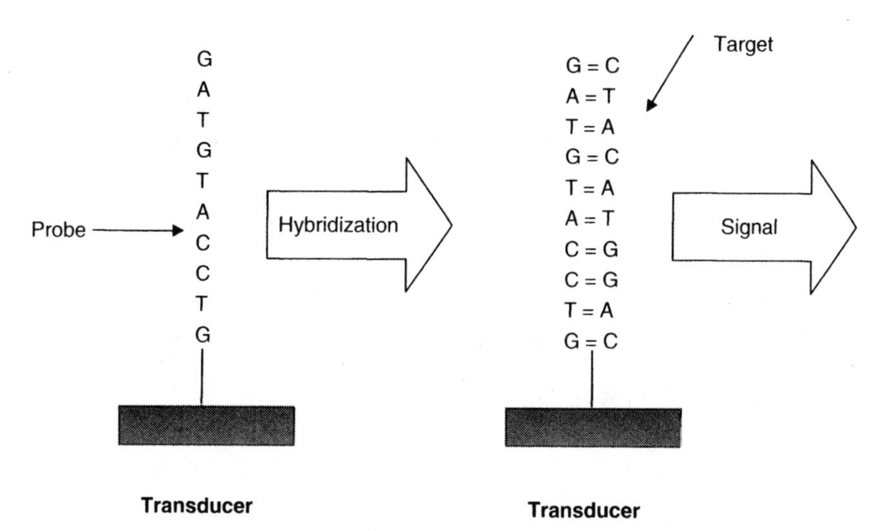

Figure 6.16 Steps involved in the detection of a specific DNA sequence using an electrochemical DNA hybridization biosensor. (Reproduced with permission from Ref. 71.)

chemical parameters (e.g., capacitance or conductivity). Control of the probe immobilization (e.g., linking chemistry, surface coverage) is essential for assuring high reactivity, orientation and/or accessibility, and stability of the surface-bound probe, as well as for avoiding non-specific binding/adsorption events. Control of the hybridization conditions (e.g., ionic strength, temperature, time) is also crucial for attaining high sensitivity and selectivity (including the detection of point mutations).

6.1.2.2.2 Electrical Transduction of DNA Hybridization Several studies have demonstrated the utility of electroactive indicators for monitoring the hybridization event (67). Such redox-active compounds have a much larger affinity for the resulting duplex (compared to their affinity to the probe alone). Their association with the surface duplex thus results in an increased electrochemical response. Very successful has been the use of a threading intercalator ferrocenyl naphthalene diimide (FND) (72), which binds to the DNA duplex more tightly than do the usual intercalators and displays a negligible affinity to the single-stranded probe (Fig. 6.17). It is also possible to employ metal nanoparticle labels (e.g., colloidal gold), and to quantitate them following the hybridization and acid dissolution by a highly sensitive electrochemical stripping protocol [73].

The use of enzyme labels to generate electrical signals also offers great promise for ultrasensitive electrochemical detection of DNA hybridization. This can be accomplished by combining the hybridization step with an electrochemical measurement of the product of the enzymatic reaction. The

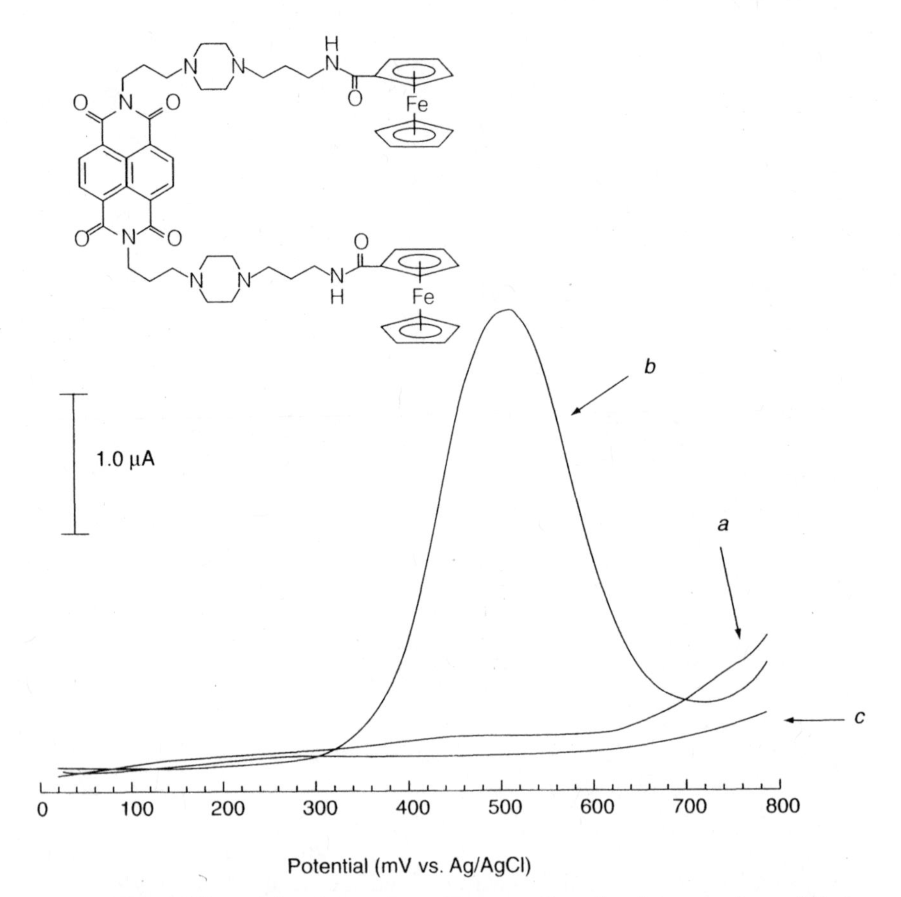

Figure 6.17 Differential-pulse voltammograms for the ferrocenyl naphthalene diimide indicator at the dT_{20}-modified electrode before (curve *a*) and after (curve *b*) hybridization with dA_{20}. Also shown is the chemical structure of the indicator. (Reproduced with permission from Ref. 72.)

potential of enzyme labels for electrical detection of DNA hybridization was demonstrated using horseradish peroxidase [74] or alkaline phosphatase [75]. Such enzymatic amplification facilitated measurements down to the zmol (3000 copies) level [74].

It is also possible to exploit different rates of electron transfer through ss- and ds-DNA for probing hybridization (including mutation detection) via the perturbation in charge migration through DNA. Barton's group found that such charge transport is very sensitive to the DNA structure and perturbations in the structure and exploited such DNA-mediated charge transport chemistry for detecting single-base mutations and DNA damage [76].

Increased attention has been given to new indicator-free electrochemical detection schemes that greatly simplify the sensing protocol. Such direct, label-free, electrical detection of DNA hybridization can be accomplished by monitoring changes in the conductivity of conducting polymer molecular interfaces, such as by using DNA-substituted or doped polypyrrole films [77]. It is also possible to exploit changes in the intrinsic electroactivity of DNA accrued from the hybridization event [78]. Among the four nucleic acids bases, the guanine moiety is most easily oxidized and is most suitable for such label-free hybridization detection [68]. A greatly amplified guanine signal, and hence hybridization response, can be obtained by using the electrocatalytic action of a $Ru(bpy)_3^{2+}$ redox mediator [78]. Such mediated guanine oxidation is illustrated in Figure 6.18. Ultimately, these developments will lead to the introduction of miniaturized (on-chip) sensor arrays, containing numerous microelectrodes (each coated with a different oligonucleotide probe) for the simultaneous hybridization detection of multiple DNA sequences. The new gene chips would integrate a microfluidic network, essential for performing all the steps of the bioassay (see Section 6.3.2), and would thus address the growing demands for shrinking DNA diagnostics, in accordance to the market needs in the twenty-first century.

6.1.2.2.3 Other Electrochemical DNA Biosensors Other modes of DNA interactions (besides base-pair recognition) can be used for the development of electrochemical DNA biosensors. In particular, dsDNA-modified electrodes can be designed for detecting small molecules (e.g., drugs or carcinogens) interacting with the immobilized nucleic acid layer (80,81). The intercalative accumulation of these molecules onto the surface-confined DNA layer can be used to measure them at trace levels, in a manner analogous to the preconcentration/voltammetric schemes described in Section 4.5.3.4. In addition, the sensitivity of the intrinsic redox signals of DNA to its structure and conformation offers considerable opportunities for nucleic acid research, including the sensing of DNA damage (82). Damage to DNA in the cells may result in

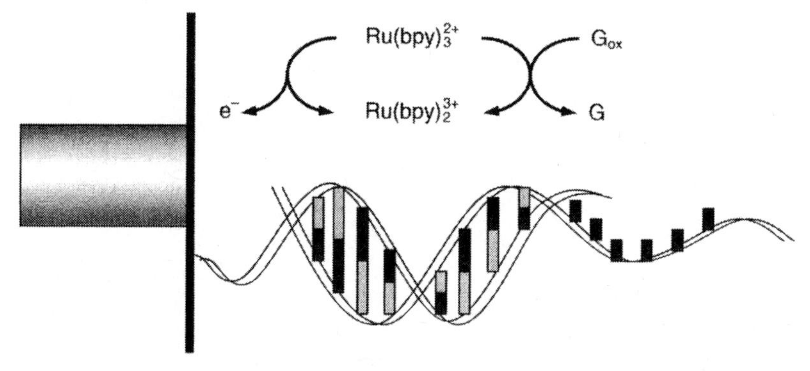

Figure 6.18 Schematic representation of guanine oxidation mediated by a ruthenium complex. (Reproduced with permission from Ref. 79.)

a serious disturbance of cell life, including apoptosis or malignant cell transformation. Various studies have shown that pulse polarographic and adsorptive stripping analyses can detect small damage to DNA induced by various chemical agents, enzymatic digestion, or ionizing radiation (83,84). Such electrical detection of DNA damage reflects the fact that the electrochemical response of DNA is strongly dependent on the DNA structure. In addition to monitoring changes in the intrinsic faradaic (e.g., guanine oxidation) response or tensammetric signals of DNA at solid or mercury electrodes, such studies examine response of electroactive damaging agents (e.g., carcinogens) interacting with DNA, often in connection to a DNA-modified electrode (82–84).

6.1.2.3 Receptor-Based Sensors Another promising and new sensing avenue is the use of chemoreceptors as biological recognition elements. Receptors are protein molecules embedded in the cellular membrane, and specifically bind to target analytes. The receptor–analyte (host–guest) binding can trigger specific cellular events, such as modulation of the membrane permeability, or activate certain enzymes, which translate the chemical interaction to electrical signals. For example, ion channel sensors, utilizing receptors in a bilayer lipid membrane, couple the specific binding process with intense signal amplification (85–87). The latter is attributed to the opening/closing switching of the ion flux through the membrane (Fig. 6.19). A single selective binding event between the membrane receptor and the target analyte can thus result in an increase of the transmembrane conduction that involves thousands of ions. Unlike most antibody bindings (aimed at specific substances), receptors tend to bind to classes of substances (possessing common chemical properties that dictate the binding affinity). Accordingly, receptor-based biosensors are usually class-specific devices.

Instead of isolating, stabilizing, and immobilizing chemoreceptors onto electrodes, it is possible to use intact biological sensing structures for determining relevant chemical stimulants (88,89). This novel concept was illustrated with antennule structures of the blue crab. Such structures are part of the crab food-

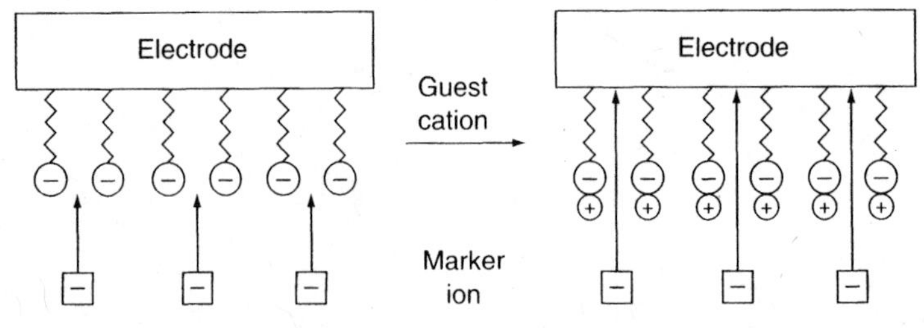

Figure 6.19 Schematic representation of the ion permeability modulation for cation-responsive voltammetric sensors based on negatively charged lipid membranes. Complexation of the guest cation to the phospholipid receptors causes an increase of the permeation for the anionic marker ion. (Reproduced with permission from Ref. 85.)

locating system, and thus can be exploited for the determination of amino acids. Similarly, various drugs can be monitored with respect to their stimulation of nerve fibers in the crayfish walking leg. A flow cell based on such a neuronal sensor is shown in Figure 6.20. Such a sensor responds to stimulant compounds at extremely low levels (down to 10^{-15} M), with very short response times. The relationship between the response frequency (R) and the stimulant concentration (C) is given by

$$R = \frac{R_{max}}{1 + (K/C)^n} \tag{6.14}$$

where R_{max} is the maximum response frequency, n is a cooperativity factor between receptors, and K is a constant.

In addition to the use of bioreceptors, it is possible to design artificial molecules that mimic bioreceptor functions (87). Such artificial receptors (hosts) can be tailored for a wide range of guest stimulants. For example, cyclodextrin derivatives have been used to provide a shape discrimination effect in connection with ion channel sensors (90). The receptors are incorporated within artificial lipid membranes, prepared by the Langmuir–Blodgett (LB) deposition method on the transducer surface. The LB method (which involves a transfer of a monomolecular film from an air–water interface onto the electrode surface) results in thin organic films that can be organized into multilayer molecular assemblies one monolayer at a time. Artificial sensor arrays (emulating biological sensory systems, e.g., the human nose) are also being explored in various laboratories (see Ref. 91 and Section 6.4). High

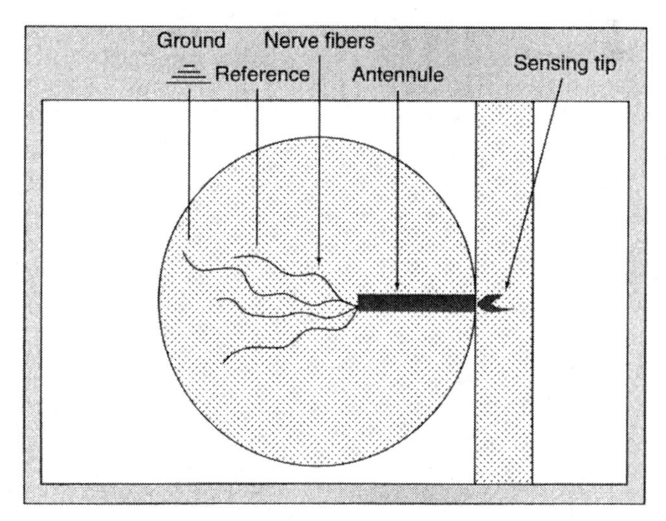

Figure 6.20 Neuronal sensing—top view of a flow cell with mounted antennule and the various electrode connections. (Reproduced with permission from Ref. 88.)

sensitivity and selectivity can be achieved also by using the receptor recognition process as an in situ preconcentration step (92).

6.1.2.4 Electrochemical Sensors Based on Molecularly Imprinted Polymers

Molecular imprinting is an attractive approach to mimic natural molecular recognition by preparing synthetic recognition sites with predetermined selectivity for various target analytes (93). In this method, the target analyte is engaged as a template molecule that binds functional monomers by covalent or noncovalent bonding during the polymerization process. The resulting macroporous polymers contain recognition sites that, because to their shape and arrangement of functional groups, possess high affinity for the print molecule. The selectivity and affinities acquired from the molecular imprinting process approach those of biological recognition elements, such as antibodies. The high stability is coupled to long-term stability and resistance to harsh environments. Molecule-imprinted polymers have been combined with a wide range of amperometric and potentiometric transducers (94–96). A recent review covers electrochemical sensors based on molecularly imprinted polymers (97). Additional information on molecule-imprinted sensors using conducting-polymer or solgel materials is given in Chapter 4.

6.2 GAS SENSORS

Real-time monitoring of gases, such as carbon dioxide, oxygen, and ammonia, is of great importance in many practical environmental, clinical, or industrial situations. Gas-sensing electrodes are highly selective devices for measuring dissolved gases. They are reliable and simple, exhibit excellent selectivity, but tend to have relatively slow response times (particularly as the limit of detection is approached).

Gas sensors usually incorporate a conventional ion-selective electrode surrounded by a thin film of an intermediate electrolyte solution and enclosed by a gas-permeable membrane. An internal reference electrode is usually included, so that the sensor represents a complete electrochemical cell. The gas (of interest) in the sample solution diffuses through the membrane and comes to equilibrium with the internal electrolyte solution. In the internal compartment, between the membrane and the ion-selective electrode, the gas undergoes a chemical reaction, consuming or forming an ion to be detected by the ion-selective electrode. (Protonation equilibria in conjunction with a pH electrode are most common.) Since the local activity of this ion is proportional to the amount of gas dissolved in the sample, the electrode response is directly related to the concentration of the gas in the sample. The response is usually linear over a range of typically four orders of magnitude; the upper limit is determined by the concentration of the inner electrolyte solution. The permeable membrane is the key to the electrode's gas selectivity. Two types of polymeric material, microporous and homogeneous, are used to form the gas-permeable membrane. Typically, such membranes are 0.01–0.1 mm in thickness. Such hydrophobic membranes are impermeable to water or ions. Hence, gas-sensing probes exhibit excellent selectivity, compared with many ion-selective electrodes. Besides the membrane, the response characteristics are often affected by the composition of internal solution and the variables of geometry (98). Amperometric gas sensors based on different configurations have also been developed. These often rely on fixing a high-surface-area electrode on a solid polymer electrolyte (facing an internal electrolyte solution, containing the reference and counter electrodes). Such amperometric gas sensors have been reviewed (99,100). One-dimensional nanomaterials, such as nanowires or carbon nanotubes, have also received considerable attention for conductivity monitoring of gases (101). Because of the high-surface:volume ratio of these nanostructures, their electronic conductance is strongly influenced by minor surface perturbations (such as those associated with the adsorption of gases).

6.2.1 Carbon Dioxide Sensors

Carbon dioxide devices were originally developed by Severinghaus and Bradley (102) to measure the partial pressure of carbon dioxide in blood. This electrode, still in use today (in various automated systems for blood gas analysis), consists of an ordinary glass pH electrode covered by a carbon dioxide membrane, usually silicone, with an electrolyte (sodium bicarbonate–sodium chloride) solution entrapped between (Fig. 6.21). When carbon dioxide from the outer sample diffuses through the semipermeable membrane, it lowers the pH of the inner solution:

$$CO_2 + H_2O \rightarrow HCO_3^- + H^+ \tag{6.15}$$

Such changes in pH are sensed by the inner glass electrode. The overall cell potential is thus determined by the carbon dioxide concentration in the sample:

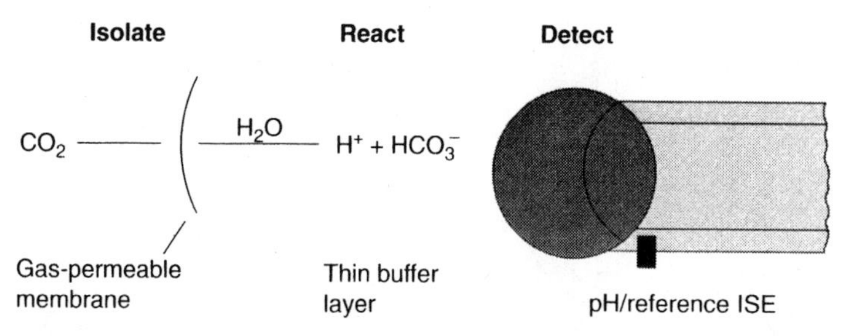

Figure 6.21 Schematic representation of the potentiometric sensor for carbon dioxide. (Reproduced with permission from Ref. 103.)

$$E = K + (RT/F)\ln[CO_2] \tag{6.16}$$

Such a Nernstian response of 59-mV/decade changes in concentration is commonly observed (at 25°C). Relation to the partial pressure carbon dioxide is accomplished by the use of Henry's law. A catheter sensor configuration has been developed for in vivo monitoring of blood carbon dioxide (104).

By using different membranes, it is possible to obtain potentiometric sensors for gases such as sulfur dioxide or nitrogen dioxide. Such sensors employ similar (acid–base) or other equilibrium processes. These devices, along with their equilibrium processes and internal electrodes, are summarized in Table 6.2. Membrane coverage of other ion-selective electrodes (e.g., chloride) can be used for the sensing of other gases (e.g., chlorine).

6.2.2 Oxygen Electrodes

While most gas sensors rely on potentiometric detection, the important oxygen probe is based on amperometric measurements. In particular, membrane-covered oxygen probes based on the design of Clark et al. (105) have found acceptance for many applications. The sensor is based on a pair of electrodes immersed in an electrolyte solution and separated from the test solution by a gas-permeable hydrophobic membrane (Fig. 6.22). The membrane is usually

TABLE 6.2 Potentiometric Gas Sensors

Target Gas	Equilibrium Process	Sensing Electrode
CO_2	$CO_2 + H_2O \rightleftharpoons HCO_3^- + H^+$	H^+, CO_3^{2-}
NO_2	$2NO_2 + H_2O \rightleftharpoons NO_3^- + NO_2^- + 2H^+$	H^+
SO_2	$SO_2 + H_2O \rightleftharpoons HSO_3^- + H^+$	H^+
H_2S	$H_2S \rightleftharpoons 2H^+ + S^{2-}$	S^{2-}, H^+
HF	$HF \rightleftharpoons H^+ + F^-$	F^-, H^+

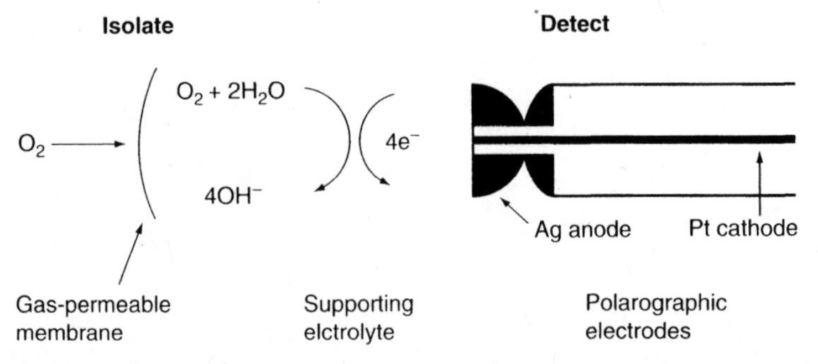

Figure 6.22 Membrane-covered oxygen probe based on the Clark electrode. (Reproduced with permission from Ref. 103.)

made of Teflon, silicon rubber, or polyethylene, while the electrolyte is a solution of potassium chloride and buffer. Oxygen diffuses through the membrane and is reduced at the surface of the sensing electrode. The resulting electrolytic current is proportional to the rate of diffusion of oxygen to the cathode, and hence to the partial pressure of oxygen in the sample. Such an electrode thus displays a linear response (as compared to the logarithmic dependence of most gas sensors, discussed in Section 6.2.1). The actual potential applied at the cathode (with respect to the anode/reference electrode) depends on the particular design. Cathodes made of platinum, gold, or silver are commonly incorporated in different commercial probes. The applied potential usually maintains the cathode on the diffusion-limited plateau region for the oxygen reduction process. Periodic calibration is desired for addressing slow drifts. This is usually accomplished by exposure to samples with known oxygen content, for instance, with air assumed to be 20.93% O_2. The response time of the electrode is generally larger when changing from a high P_{O_2} to a low P_{O_2}, compared with a change in the opposite direction. The applications of various oxygen sensors have been reviewed (106).

Membraneless oxygen sensors based on solid-state technology have also been reported. For example, coverage of a Y_2O_3-doped ZrO_2 disk with porous platinum electrodes results in a selective sensor, based on the coupling of the oxygen reduction process and the preferential transport of the oxide ion product through vacancies in the doped crystal (107). For this purpose, one of the platinum electrodes is exposed to the unknown gas, while the second one is exposed to the reference gas. Such potentiometric sensors commonly operate at high temperatures, and are widely used in the automotive industry for controlling the ratio of air/fuel (with an annual worldwide market exceeding $150 million).

Other useful gas sensors include the potentiometric ammonia (108) or hydrogen cyanide probes (109), and amperometric carbon monoxide (110), nitrogen dioxide (111), and ozone or aldehyde (100) devices. The latter rely on the use of gold–Nafion electrodes prepared by depositing gold onto a solid polymer electrolyte. The hydrogen cyanide probe is an example of a modern device that relies on changes in the conductivity of electropolymerized film (polyaniline) in the presence of a given gas.

6.3 SOLID-STATE DEVICES

6.3.1 Ion-Selective Field Effect Transistors

The integration of chemically sensitive membranes with solid-state electronics has led to the evolution of miniaturized, mass-produced potentiometric probes known as *ion-selective field effect transistors* (ISFETs). The development of ISFETs is considered as a logical extension of coated-wire electrodes (described in Section 5.2.4). The construction of ISFETs is based on the technology used to fabricate microelectronic chips. Ion-selective field effect transistor incorporates the ion-sensing membrane directly on the gate area of a field effect transistor (FET) (Fig. 6.23). The FET is a solid-state device that exhibits high-input impedance and low-output impedance and therefore is capable of monitoring charge buildup on the ion-sensing membrane. As the charge density on this membrane changes because of interaction with the ions in solution, a drain current is flowing between the source and the drain of the transistor. The increased voltage needed to bring the current back to its initial value represents the response. (This is commonly accomplished by placing the ISFET in a feedback loop.) From the standpoint of change in drain current as a result of change in activity of the ion of interest, the ISFET response is governed by the same Nernstian relationship (and the selectivity limitation) that characterized conventional ion-selective electrodes.

Such sensors that utilize solid-state electronics have significant advantages. The actual sensing area is very small. Hence, a single miniaturized solid-state chip could contain multiple gates and be used to sense several ions simultaneously. Other advantages include the in situ impedance transformation and the ability for temperature and noise compensation. While the concept of ISFET is thus very exciting and intriguing, problems with stability and encapsulation still need to be solved before such devices reach the truly practical stage. One problem is the detachment of PVC-type ion-sensing membranes from the gates of FETs. This problem can be minimized by suspending a polyimide mesh over the gate (112); the polymer film thus becomes anchored in place by the mesh.

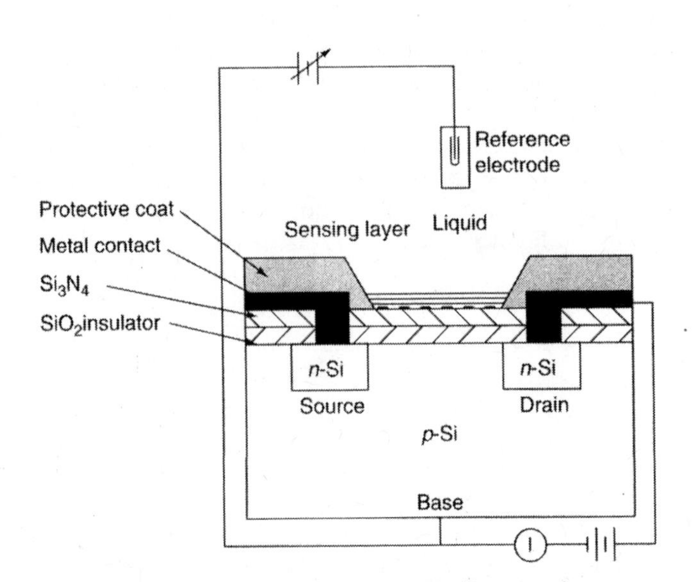

Figure 6.23 An ion-selective field effect transistor.

The coating on the gate is key to the analytical chemistry that the ISFET can perform. Ion-selective field effect transistors based on various ion-responsive layers have been developed. Among these are a sodium ISFET based on the synthetic sodium carrier ETH 227, an ammonium ISFET utilizing monactin–nonactin (113), and a chloride ISFET prepared by laying a membrane of methyltridodecylammonium chloride of a silicon nitride gate (114). Ion-selective field effect transistors that are not covered with an ion-responsive membrane can be used directly as pH sensors. The silicon nitride coating on the transistor is itself sensitive to hydrogen ions (by its own surface properties), and develops phase boundary potentials proportional to the logarithm of the hydrogen ion activity in the contacting solution. The ability of sensing several ions was illustrated using a quadruple-function ISFET probe that simultaneously monitored potassium, sodium, calcium, and pH in whole blood samples (115). Ion-selective field effect transistors can be combined with various biological agents, such as enzymes and antigens, to form effective biosensors. The biological recognition process results in modulation of the gate voltage, and thus controls the drain current. For example, an enzyme field effect transistor (ENFET) was developed for continuous monitoring of glucose in body fluids (116). The theory and mode of operation of ISFETs have been reviewed (117).

6.3.2 Microfabrication of Solid-State Sensor Assemblies

Other miniaturized solid-state sensors can be fabricated by coupling microelectronics and chemically sensitive layers. In particular, Wrighton and coworkers (118) fabricated diode and transistor structures by combining a conducting polymer with lithographically defined interdigitated microarray electrodes. Such devices have been responsive to redox species such as oxygen or hydrogen. Interesting biosensing applications of this molecular electronic switching device involved the addition of an enzyme layer (119,120). Changes in the conductivity of the polymer resulted from pH changes (associated with the enzymatic reaction), have thus been exploited for monitoring the corresponding substrate (e.g., see Fig. 6.24). Miniaturized and disposable amperometric biosensors can be achieved by coupling microfabricated oxygen electrodes with various biocomponents (121).

6.3.3 Microfabrication Techniques

Microfabrication technology has made a considerable impact on the miniaturization of electrochemical sensors and systems. Such technology allows replacement of traditional bulky electrodes and "beaker"-type cells with mass-producible, easy-to-use sensor strips. These strips can be considered as disposable electrochemical cells onto which the sample droplet is placed. The development of microfabricated electrochemical systems thus has the potential to revolutionize the field of electroanalytical chemistry.

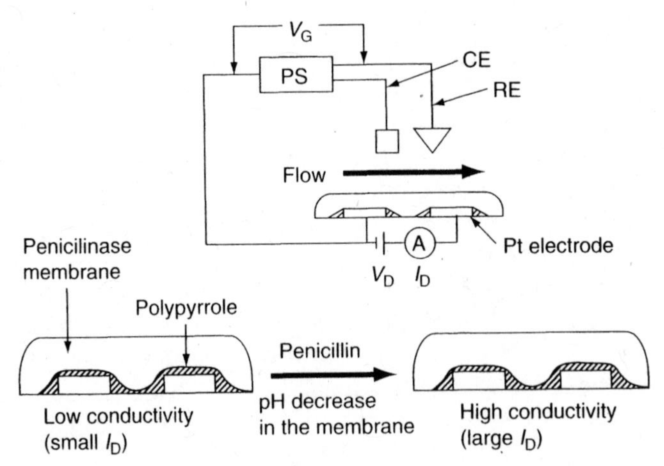

Figure 6.24 Configuration of a penicillin sensor based on a microarray electrode coated with a pH-responsive polypyrrole. (Reproduced with permission from Ref. 120.)

The thin-film lithographic approach can be used for producing small (micrometer) dimension electrodes on silicon wafers (e.g., see Fig. 4.30). This technology couples various processes (based on electronic integrated-circuit manufacturing), including vapor deposition of a thin metal film, its coverage with a UV-sensitive photoresist (by spin coating), photolithographic patterning (using a proper photomask and UV radiation), removal of the exposed photopolymerized soluble zone of the photoresist with a developer, and chemical or plasma etching (122). The resist can be rendered soluble or insoluble (i.e., positive or negative) using radiation in a developing solution, depending on the features required. Three-electrode microsystems can thus be readily prepared on a planar silicon wafer. This lithographic approach represents an attractive route for the production of sensor arrays (123). For example, Figure 6.25 displays a multiple-analyte sensor array, commonly used for point-of-care clinical assays of small blood droplets (124,125). This commercial (i-STAT Inc.) handheld system produces results on eight commonly requested clinical tests in less than 2 min. Lithographic routes have also been employed for fabricating microcavity devices for ultra-small-volume assays (126).

Large-scale sensor fabrication can be accomplished not only by lithographic techniques but also by using modern screen-printing (thick-film) processes (127,128). The screen-printing technology relies on printing patterns of conductor and insulators onto the surface of planar (plastic or ceramic) substrates. Various conducting and insulating ink materials are available for this task. The screen-printing process involves several steps (as illustrated in Fig. 6.26 for the fabrication of carbon electrodes), including placement of the ink onto a screen or stencil containing patterned opening, followed by forcing it through the screen with the aid of a squeegee, and drying and/or curing the printed patterns. Such a process yields mass-producible (uniform and disposable) elec-

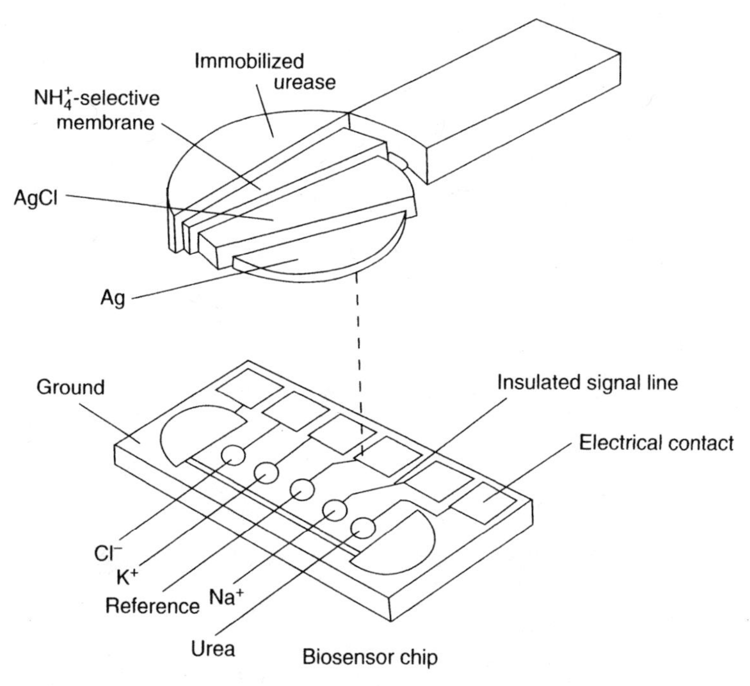

Figure 6.25 A silicon-based sensor array for monitoring various blood electrolytes, gases, and metabolites. (Courtesy of i-STAT Co.)

trodes of different shapes or sizes, similar to the glucose strip shown in Figure 6.9. The electrochemical reactivity and overall performance of screen-printed electrodes are dependent on the composition of the ink employed and the printing and curing conditions (e.g., pressure, temperature). Disposable potentiometric sensors can be fabricated by combination of ion-selective polymeric membranes with dry reagent films. Such disposable ion-selective electrode slides require microliter (10–50-μL) sample volumes, and are ideally suited for various decentralized applications. Mass-produced potentiometric sensor arrays are also being developed for use in future high-speed clinical analyzers (129). These are being combined with advanced materials (e.g., hydrogels) that circumvent the need for internal filling solutions (common to ISE sensors; see Chapter 5). The screen-printing technology requires lower capital and production costs than does the thin-film lithographic approach, but is limited to electrode structures larger than 100μm. It is also possible to fabricate electrochemical devices, combing the thin- and thick-film processes.

Commercial thin- and thick-film biosesnors commonly rely on ink-jet noncontact localization of the regent/recognition layer. Such ink-jet printing allows fluid to be deposited with high speed, extremely low volume, and great accuracy.

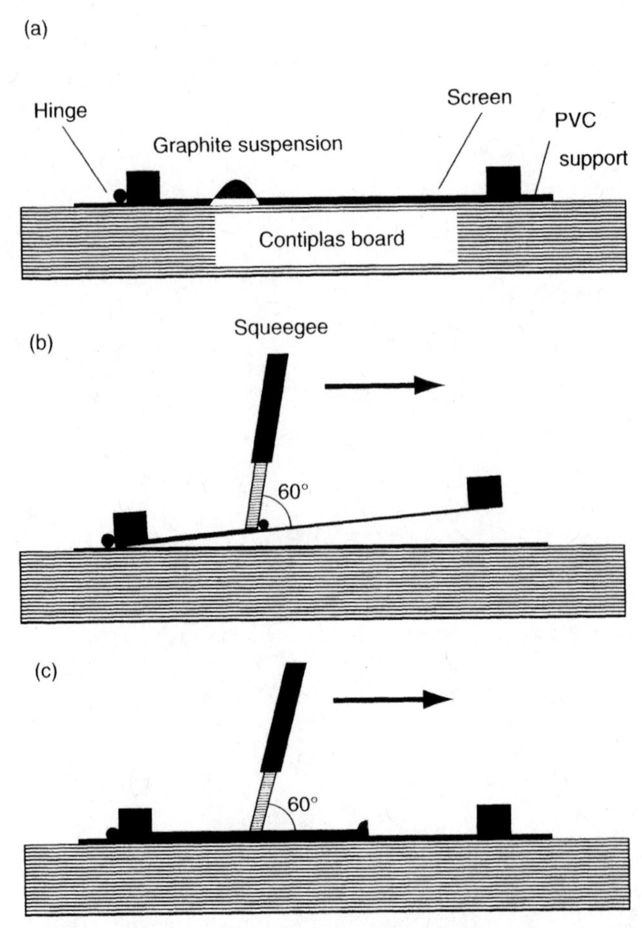

Figure 6.26 Steps involved in the screen printing process: (a) deposit the graphite suspension onto the screen; (b) load the screen mesh with the graphite; (c) force the graphite onto the substrate. (Reproduced with permission from Ref. 128.)

6.3.4 Micromachined Analytical Microsystems

Modern fabrication technologies can be used to produce complete miniaturized analytical systems integrating glass or polymer microstructures, such as microchannels, mixing chambers, microreactors, or valves) (130–132). Because of their fluid manipulation capability, such micromachined systems hold great promise for performing all the steps of a chemical or biological assay (including the sample preparatory steps, analytical reactions, separation, and detection) on a single-microchip platform, and offer greatly improved efficiency with respect to sample size, reagent/solvent consumption, and response time. For obvious reasons, such devices are referred to as "lab-on-a-chip" devices or micro–total analytical systems (μTAS). The advantages of such analytical microsystems, including speed, high performance, integration, versatility, negligible sample/reagent consumption, miniaturization, and automation, have been well documented (131,132). Precise fluid control is accomplished by regulating the applied potentials at the terminus of each channel of the microchip (133). Such electrosmotic flow obviates the need for pumps or valves. The channel network of these microchips includes mixing tees and cross intersections for mixing reagents and injecting samples with high reproducibility (e.g., see Fig. 6.27). Such on-chip integration of the sample manipulations and detection should find use in numerous analytical applications, and will potentially enable the laboratory to be transported to the sample. A range of assays has already been adapted to the microchip format.

Electrochemical detection has attracted considerable interest for miniaturized analytical systems (134,135). The extremely small dimensions of electrochemical detectors, coupled with their remarkable sensitivity, compatibility

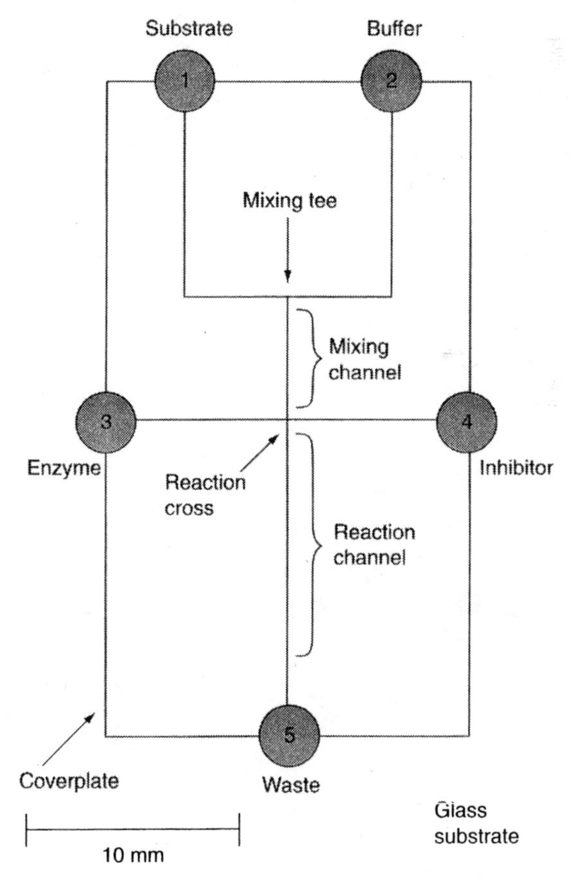

Figure 6.27 Schematic representation of the channels terminating at reservoirs containing the indicated solutions. (Reproduced with permission from Ref. 133.)

with microfabrication technologies, and low cost, make them very suitable for adaptation for "lab-on-a-chip" analytical microsystems. The electrochemical route allows on-chip integration of the control instrumentation to produce self-contained, truly portable microanalytical systems. Photolithographic processes have been used to incorporate the high-voltage electrodes (driving the fluid movement) with the electrochemical detection electrodes directly onto the microchip platform (Fig. 6.28). Microsystems relying on electrosmotic flow require proper attention to the decoupling of the detector potential from the high voltage used to control the microfluidic. Particularly powerful is the use of electrochemical detectors for monitoring on-chip electric-field-driven separations (see Section 4.6). Self-contained disposable DNA biochips, combining electrochemical detection with electrochemical pumping, sample preparation, and DNA amplification, have been developed and successfully applied for whole blood analysis (137).

6.4 SENSOR ARRAYS

So far we have discussed the one-sensor/one-analyte approach based on highly selective recognition elements. However, arrays of independent electrodes can offer much more analytical information and thus hold a great potential for many practical applications (138). These include the development of "intelligent sensing systems" capable of responding to changes in the chemical environment of the array.

The use of multielectrode arrays takes advantage of the partial selectivity of an individual electrode, by combining several electrodes and examining the

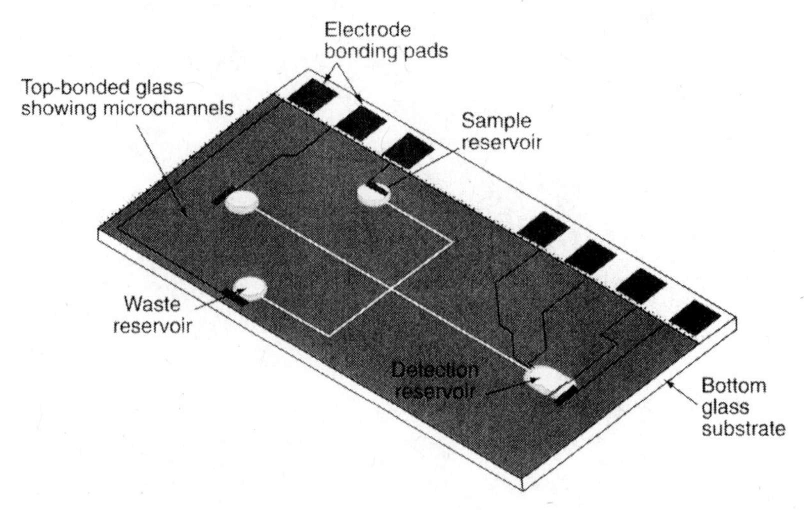

Figure 6.28 Fully integrated on-chip electrochemical detection for capillary electrophoresis in a microfabricated device. (Reproduced with permission from Ref. 136.)

relative responses of all the sensors together. The array's response of each analyte thus corresponds to a fingerprint pattern (e.g., see Fig. 6.29). In addition, the coupling of multielectrode arrays with a chemometric (multivariate calibration) approach allows analysis of a mixture of analytes. Three calibration techniques, partial least squares (PLS), principal-component regression (PCR), and multiple linear regression (MLR), are particularly useful for this task. These techniques take advantage of the response pattern produced by the sensor array for a given analyte to both identify and quantitate that component. Such use of mathematics thus provides the necessary selective response from nonselective multivariate data. High stability, rather than selectivity, is the primary concern in the operation of electrode arrays.

Various types of multielectrode arrays can be employed. For example, potentiometric electrode arrays exploit the fact that ion-selective electrodes respond to some degree to a range of ions (140–142). The first potentiometric array was described by Otto and Thomas (140). Diamond and coworkers have illustrated the utility of an array consisting of three highly selective electrodes along with a sparingly selective one (142). Arrays of highly selective potentiometric electrodes can also be useful when high-speed analysis is concerned (143). Arrays of voltammetric electrodes can be based on the use of different electrode materials (144) or catalytic surface modifiers (139) (with different voltammetric characteristics), on the use of partially selective coated electrodes (each covered with a different permselective film) (145), or with the use of different operating potentials or surface pretreatments (146,147). For example, the use of four uncoated electrodes held at different applied potentials was shown extremely useful for detecting 22 organic vapors in connection to different pretreatment of the gas stream (146). Microlithographic

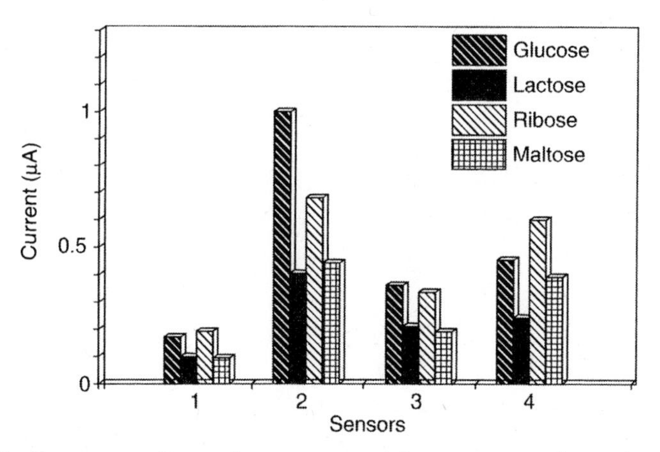

Figure 6.29 Response pattern of an amperometric sensor array for various carbohydrates. The array consisted of carbon paste electrodes doped with CoO (1), Cu_2O (2), NiO (3), and RuO_2 (4). (Reproduced with permission from Ref. 139.)

techniques are often used for the construction of such amperometric array electrodes, with multielectrode potentiostats controlling the potentials of the individual electrodes. A single (common) reference electrode is normally used with these arrays. Novel arrays of broadly responsive chemically diverse polymeric chemoresistors or metal oxide sensors have been used, in connection with computer-assisted pattern recognition algorithms for monitoring the flavor of beers (148) or for detecting various odorants (149,150). Changes in the resistivity of a series of polypyrrole-based conducting polymers, on the adsorption of different volatile compounds, have been particularly useful for creating the response patterns (150). The different, yet partially overlapping, signals are achieved by preparing the individual polymers from modified monomers units or with different counter ions (dopants). Such arrays serve as electronic analogs for the human nose. Indeed, the remarkable performance of biological olfactory systems in odor detection has served as an inspiration in the development of sensor arrays. Such biological systems rely on cross-reactive receptors that respond to many odors, generating unique response patterns (that serve as "fingerprints" for each odor). The information flow in "electronic nose" sensor arrays is displayed in Figure 6.30. Practical applications of electrode arrays have been facilitated by the availability of inexpen-

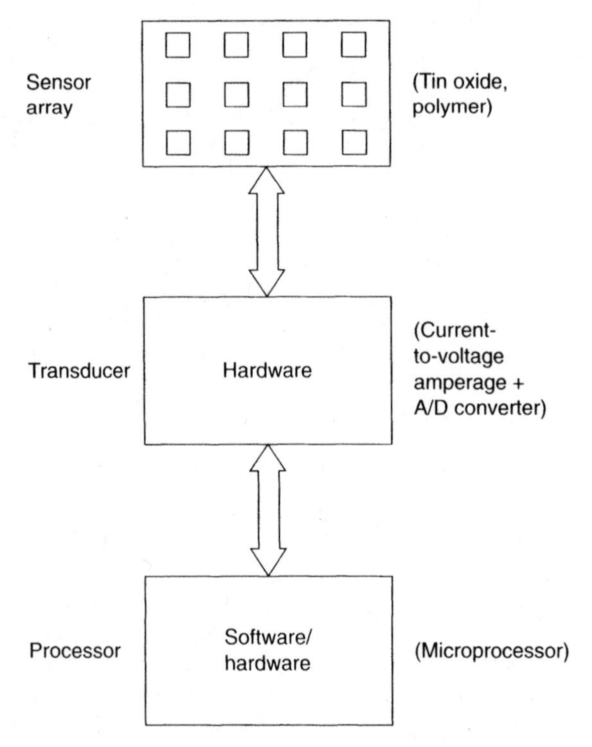

Figure 6.30 The flow of information in the array-based artificial nose. (Reproduced with permission from Ref. 149.)

sive multichannel data acquisition cards for personal computers. The development of sensor arrays has been reviewed by Diamond (151). New advances are expected in the near future based on the development of user-friendly software, of new statistical tools, and of novel sensor fabrication technology.

EXAMPLES

Example 6.1 Chronoamperogram *a* was obtained for the biosensing of glucose in whole blood. Subsequent standard additions of 1×10^{-3} M glucose yielded chronoamperograms *b*–*d*. Find the concentration of glucose in the sample.

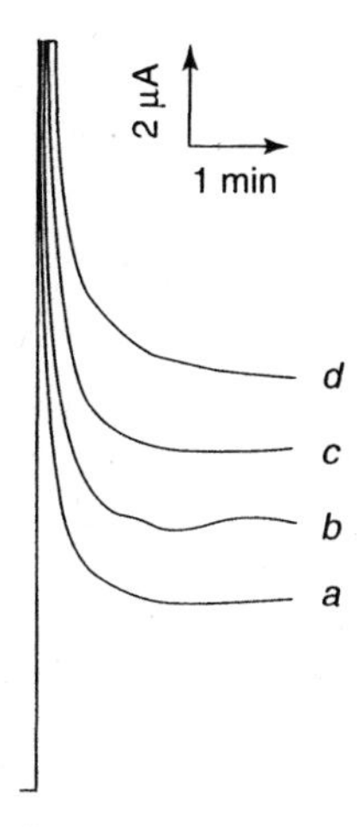

Solution The resulting current transients (sampled after 2 min) lead to the following standard addition plot:

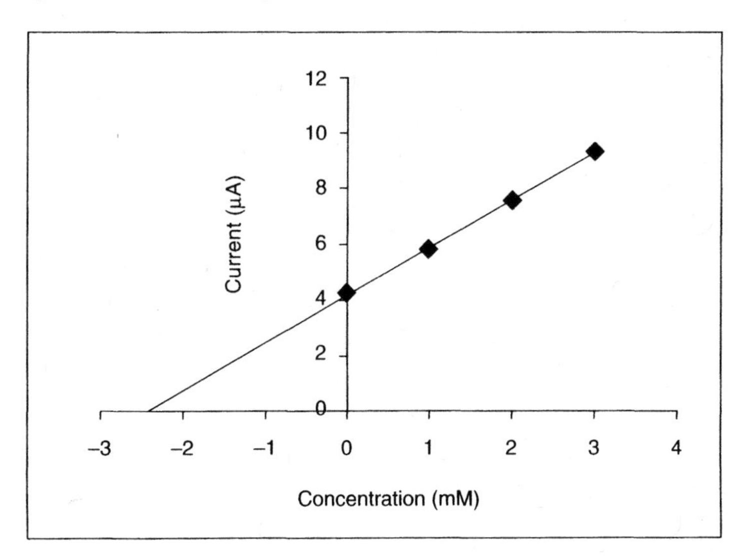

From this plot, a glucose concentration of 2.4 mM can be obtained for the sample.

Example 6.2 The following standard addition plot was obtained for a competitive electrochemical enzyme immunoassay of the pesticide 2,4D. A groundwater sample (diluted 1 : 20 fold) was subsequently assayed by the same protocol to yield a current signal of 65 nA. Calculate the concentration of 2,4D in the original sample.

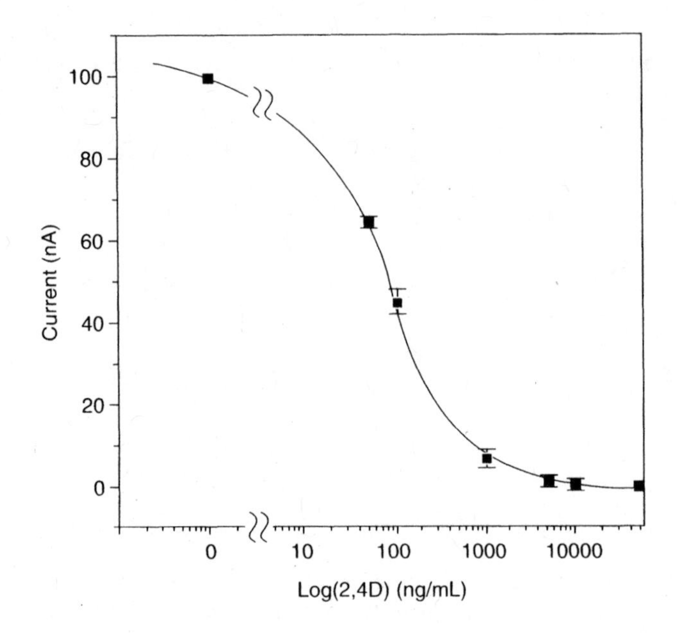

Solution The response of the diluted sample corresponds to 50 ng/mL 2,4D. Considering the 20-fold dilution, the original sample concentration corresponds to 1.0 μg/mL.

PROBLEMS

6.1 Describe different schemes for immobilizing enzymes onto electrode transducers.

6.2 How would you extend the linear range of calibration plots based on the use of enzyme electrodes?

6.3 Describe a biosensing protocol for detecting mutations in DNA samples.

6.4 Suggest an enzyme-electrode-based procedure for detecting organophosphate pesticides.

6.5 Describe various routes for facilitating the electrical communication between the redox center of glucose oxidase and electrode surfaces.

6.6 Describe the major problems encountered in the detection of the NADH product of dehydrogenase-based amperometric biosensors. Discuss a common approach to circumvent these problems.

6.7 Explain how surface chemistry and coverage affect the success of electrical DNA-hybridization detection schemes.

6.8 What are the improvements associated with performing analytical procedures on a microchip platform?

REFERENCES

1. Sittamplam, G.; Wilson, G. S., *J. Chem. Educ.* **59**, 70 (1982).
2. Willner, I.; Katz, E.; Willner, B., *Electroanalysis* **9**, 965 (1997).
3. Gorton, L., *Electroanalysis* **7**, 23 (1995).
4. Algeret, S., *Analyst* **121**, 1751 (1996).
5. Wang, J.; Angnes, L., *Anal. Chem.* **64**, 456 (1992).
6. Bartlett, P. N.; Cooper, J. M., *J. Electroanal. Chem.* **362**, 1 (1993).
7. Gray, D. N.; Keyes, M. H.; Watson, B., *Anal. Chem.* **49**, 1067A (1977).
8. Weetall, H. H., *Anal. Chem.* **46**, 602A (1974).
9. Yang, X.; Johansson, G.; Pfeiffer, D.; Scheller, F., *Electroanalysis* **3**, 659 (1991).
10. Gorton, L.; Lindgren, A.; Larsson, T.; Muteanu, F.; Ruzgas, T.; Gazaryan, I., *Anal. Chim. Acta* **400**, 91 (1999).
11. Guindilis, A.; Atanasov, P.; Wilkins, E., *Electroanalysis* **9**, 661 (1997).
12. Habermuller, K.; Mosbach, M.; Schuhmann, W., *Fres. J. Anal. Chem.* **366**, 560 (2000).
13. Wang, J., *Electroanalysis* **13**, 983 (2001).
14. Updike, S. J.; Hicks, G. P., *Nature* **214**, 986 (1967).
15. Karayakin, A., *Electroanalysis* **13**, 813 (2001).
16. Wang, J.; Liu, J.; Fang, L.; Chen, L., *Anal. Chem.* **66**, 3600 (1994).
17. Marcus, R. A.; Sutin, N., *Biochim. Biophys. Acta* **811**, 265 (1985).
18. Schlapfer, P.; Mindt, W.; Racine, P., *Clin. Chim. Acta* **57**, 283 (1974).
19. Frew, J. E.; Hill, H. A. O., *Anal. Chem.* **59**, 933A (1987).
20. Green, M. J.; Hilditch, P. I., *Anal. Proceed.* **28**, 374 (1991).
21. Newman, J.; Turner, A. P. F., *Biosens. Bioelectron.* **20**, 2435 (2005).
22. Vreeke, M.; Maidan, R.; Heller, A. J., *Anal. Chem.* **64**, 3085 (1992).
23. Patolsky, F.; Weizmann, Y.; Willner, I., *Angew Chem. Int. Ed.* **43**, 2113 (2004).
24. Xiao, Y.; Patolsky, F.; Katz, E.; Hainfeld, J. F.; Willner, I., *Science* **299**, 1877 (2003).
25. Degani, Y.; Heller, A., *J. Phys. Chem.* **91**, 1285 (1987).
26. Riklin, A.; Katz, E.; Willner, I.; Stocker, A.; Buckmann, A., *Nature* **367**, 672 (1995).
27. Henry, C., *Anal. Chem.* **70**, 594A (1998).

28. Pickup, J. C.; Hussain, F.; Evans, N. D.; Sachedina, N., *Biosens. Bioelectron.* **20**, 1897 (2005).
29. Chen, T.; Barton, S. C.; Binyamin, G.; Gao, Z.; Zhang, Y.; Kim, H.; Heller, A., *J. Am. Chem. Soc.* **123**, 8630 (2001).
30. Heller, A., *Phys. Chem. Chem. Phys.* **6**, 209 (2004).
31. Malinauskas, A.; Kulys, J., *Anal. Chim. Acta* **98**, 31 (1978).
32. Gorton, L., *J. Chem. Soc. Faraday Trans.* **82**, 1245 (1986).
33. Lobo, A. M.; Miranda, A.; Tunon, P., *Electroanalysis* **9**, 191 (1997).
34. Guilbault, G. G.; Montalvo, J. G., *J. Am. Chem. Soc.* **92**, 2533 (1970).
35. Hall, G. F.; Turner, A. P. F., *Anal. Lett.* **24**, 1375 (1991).
36. Thompson, H.; Rechnitz, G. A., *Anal. Chem.* **46**, 246 (1974).
37. Madaras, M. B.; Buck, R. P., *Anal. Chem.* **68**, 3832 (1996).
38. Blaedel, W. J.; Jenkins, R. A., *Anal. Chem.* **48**, 1240 (1976).
39. Pfeiffer, D.; Setz, K.; Schulmeister, T.; Scheller, F.; Lueck, H., *Biosens. Bioelectron.* **7**, 661 (1992).
40. Papariello, G. J.; Mukherji, A. K.; Shearer, C. M., *Anal. Chem.* **45**, 790 (1973).
41. Hall, G.; Best, D.; Turner, A. F., *Anal. Chim. Acta* **213**, 113 (1988).
42. Seegopaul, P.; Rechnitz, G. A., *Anal. Chem.* **56**, 852 (1984).
43. Kawashima, T.; Rechnitz, G. A., *Anal. Chim. Acta* **83**, 9 (1976).
44. Saini, S.; Hall, G.; Downs, M.; Turner, A. F., *Anal. Chim. Acta* **249**, 1 (1991).
45. Wang, J., *Talanta* **40**, 1905 (1993).
46. Besombes, J; Cosnier, S.; Labbe, P.; Reverdy, G., *Anal. Chim. Acta* **311**, 255 (1995).
47. Marty, J.; Garcia, D.; Rouillon, R., *Trends Anal. Chem.* **14**, 329 (1995).
48. Rechnitz, G. A., *Science* **214**, 287 (1981).
49. Wang, J.; Lin, M. S., *Anal. Chem.* **60**, 1545 (1988).
50. Aizawa, M.; Moricka, A.; Suzuki, S., *Anal. Chim. Acta* **115**, 61 (1980).
51. Skladal, P., *Electroanalysis* **9**, 737 (1997).
52. Rosen, I.; Rishpon, J., *J. Electroanal. Chem.* **258**, 27 (1989).
53. Kobos, R. K., *Trends Anal. Chem.* **6**, 6 (1987).
54. Kaku, S.; Nakanishi, S.; Horiguchi, K., *Anal. Chim. Acta* **225**, 283 (1989).
55. Warinske, A.; Benkert, A.; Scheller, F. W., *Fres. J. Anal. Chem.* **366**, 622 (2000).
56. John, R.; Spencer, M.; Wallace, G. G.; Smyth, M., *Anal. Chim. Acta* **249**, 381 (1991).
57. Katz, E.; Willner, I., *Electroanalysis* **15**, 913 (2003).
58. Patolsky, F.; Zheng, G.; Hayden, O.; Lakadamyali, M.; Zhuang, X.; Lieber, C. M., *Proc. Natl. Acad. Sci. USA* **104**, 14017 (2004).
59. Heineman, W. R.; Halsall, H., *Anal. Chem.* **57**, 1321A (1985).
60. Bauer, C.; Eremenko, A.; Forster, E.; Bier, F.; Makower, A.; Halsall, B.; Heineman, W.; Scheller, F., *Anal. Chem.* **68**, 2453 (1996).
61. Bagel, O.; Limoges, B.; Schollhorn, B.; Degrand, C., *Anal. Chem.* **69**, 4688 (1997).
62. Scheller, F.; Bauer, C.; Makower, A.; Wollenberger, U.; Warinske, A.; Bier, F., *Anal. Lett.* **34**, 1233 (2001).
63. Dequaire, M.; Degrand, C.; Limoges, B., *Anal. Chem.* **72**, 5521 (2000).

64. Kojima, K.; Hiratuka, A.; Suzuki, H.; Vano, K.; Ikebukuo, K.; Karube, I., *Anal. Chem.* **75**, 116 (2003).
65. Aguilar, Z. P.; Vandaveer, W. R. I. V.; Fritsch, I., *Anal. Chem.* **74**, 3321 (2002).
66a. Kawde, A.; Rodriguez, M.; Lee, T.; Wang, J., *Electrochem. Commun.* **7**, 537 (2005).
66b. Xiao, Y.; Lubin, A. A.; Heeger, A. J.; Plaxco, K. W., *Angew Chemie Int Ed* **117**, 5592 (2005).
67. Mikkelsen, S. R., *Electroanalysis* **8**, 15 (1996).
68. Palecek, E.; Fojta, M., *Anal. Chem.* **73**, 75A (2001).
69. Drummond, T.; Hill, M.; Barton, J. K., *Nature Biotechnol.* **21**, 1192 (2003).
70. Wang, J., *Anal. Chim. Acta* **469**, 63 (2002).
71. Gooding, J. J., *Electroanalysis* **14**, 1149 (2002).
72. Takenaka, S.; Yamashita, K.; Takagi, M.; Uto Y.; Kondo, H., *Anal. Chem.* **72**, 1334 (2000).
73. Wang, J., *Anal. Chim. Acta* **500**, 247 (2003).
74. Zhang, Y.; Kim, H.; Heller, A., *Anal Chem.* **75**, 3267 (2003).
75. Wang, J.; Xu, D.; Polsky, R.; Arzum, E., *Talanta* **56**, 931 (2002).
76. Boon, E. M.; Ceres, D.; Drummond, T.; Hill, M.; Barton, J. K., *Nature Biotechnol.* **18**, 1096 (2001).
77. Korri-Youssoufi, H.; Garnier, F.; Srivtava, P.; Godillot, P.; Yassar, A., *J. Am. Chem. Soc.* **119**, 7388 (1997).
78. Johnston, D. H.; Glasgow, K.; Thorp, H. H., *J. Am. Chem. Soc.* **117**, 8933 (1995).
79. Li, H.; Ng, T.; Cassell, A.; Fan, W.; Chen, H.; Ye, Q.; Koehne, J.; Meyyappan, M., *Nano. Lett.* **3**, 597 (2003).
80. Yang, M.; McGovern M.; Thompson, M., *Anal. Chim. Acta* **346**, 259 (1997).
81. Wang, J.; Rivas, G.; Luo, D; Cai, X.; Dontha, N.; Farias, P.; Shirashi, H., *Anal. Chem.* **68**, 4365 (1996).
82. Fojta, M., *Electroanalysis* **14**, 1449 (2002).
83. Fojta, M.; Palecek, E., *Anal. Chim. Acta* **342**, 1 (1997).
84. Mbindyo, J.; Zhou, L.; Zhang, Z.; Stuart, J.; Rusling, R.F., *Anal. Chem.* **72**, 2059 (2000).
85. Nagase, S.; Kataoka, M.; Naganawa, R.; Komatsu, R.; Odashimo, K.; Umezawa, Y., *Anal. Chem.* **62**, 1252 (1990).
86. Krull, I.; Nikolelis, D.; Brennan, J.; Brown, R.; Thompson, M.; Ghaemmaghami, V.; Kallury, K., *Anal. Proceed.* **26**, 370 (1991).
87. Odashima, K.; Sugawara, M.; Umezawa, Y., *Trends Anal. Chem.* **10**, 207 (1991).
88. Buch, R. M.; Rechnitz, G. A., *Anal. Chem.* **61**, 533A (1989).
89. Leech, D.; Rechnitz, G. A., *Electroanalysis* **5**, 103 (1993).
90. Odashima, K.; Kotato, M.; Sugawara, M.; Umezawa, Y., *Anal. Chem.* **65**, 927 (1993).
91. Persaud, K. C., *Anal. Proceed.* **28**, 339 (1991).
92. Wang, J.; Lin, Y.; Eremenko, A.; Kurochkin, I.; Mineyeva, M., *Anal. Chem.* **65**, 513 (1993).
93. Kriz, D.; Ramstrom, O.; Mosbach, K., *Anal. Chem.* **69**, 345A (1997).
94. Kriz, D.; Mosbach, K., *Anal. Chim. Acta* **300**, 71 (1995).
95. Hutchins, R. S.; Bachas, L. G., *Anal. Chem.* **67**, 1654 (1995).

96. Kitade, T.; Kitamura, K.; Konishi, T.; Takegami, S.; Okuno, T.; Ishikawa, M.; Wakabayashi, M.; Nishikawa, K.; Muramatsu, Y., *Anal. Chem.* **76**, 6802 (2004).
97. Piletsky, S. A.; Turner, A. P., *Electroanalysis* **14**, 317 (2002).
98. Ross, J. W.; Riseman, J.; Kruger, J., *Pure Appl. Chem.* **36**, 473 (1973).
99. Gao, Z.; Buttner, W.; Stetter, J., *Electroanalysis* **4**, 253 (1990).
100. Knake, R.; Hauser, P. C., *Anal. Chim. Acta* **500**, 145 (2003).
101. Li, J.; Lu, Y. J.; Ye, Q.; Cinke, M.; Han, J.; Meyyappan, M., *Nano Lett.* **3**, 929 (2003).
102. Severinghaus, J. W.; Bradley, A. F., *J. Appl. Physiol.* **13**, 515 (1957).
103. Czaban, J. D., *Anal. Chem.* **57**, 345A (1985).
104. Opdycke, W.; Meyerhoff, M. E., *Anal. Chem.* **58**, 950 (1986).
105. Clark, L. C.; Wolf, R.; Granger, D.; Taylor, Z., *J. Appl. Physiol.* **689** (1953).
106. Ramamoorthy, R.; Dutta, P. K.; Akbar, S. A., *J. Mater. Sci.* **38**, 4271 (2003).
107. *NASA Technical Briefs* **9**, 105 (1985).
108. Meyerhoff, M. E., *Anal. Chem.* **52**, 1532 (1980).
109. Langmaier, J.; Janata, J., *Anal. Chem.* **64**, 523 (1992).
110. Xing, X.; Liu, C. C., *Electroanalysis* **3**, 111 (1991).
111. Chang, S.; Stetter, J., *Electroanalysis* **2**, 359 (1990).
112. Blackburn, G.; Janata, J., *J. Electrochem. Soc.* **129**, 2580 (1982).
113. Oesch, V.; Caras, S.; Janata, J., *Anal. Chem.* **53**, 1983 (1981).
114. Zhukova, T. V., *Zavod. Lab.* **50**, 18 (1984); *Chem. Abstr.* **101**, 221484S (1984).
115. Sibbard, A.; Covington, A. K.; Carter, R. F., *Clin. Chem.* **30**, 135 (1984).
116. Kimura, J., *J. Electrochem. Soc.* **136**, 1744 (1989).
117. Janata, J.; Huber, R. J., *Ion-Select. Electrode. Rev.* **1**, 31 (1979).
118. Kittlesen, G.; White, H.; Wrighton, M., *J. Am. Chem. Soc.* **106**, 7389 (1984).
119. Bartlett, P. N.; Birkin, P., *Anal. Chem.* **65**, 1118 (1993).
120. Nishizawa, M.; Matsue, T.; Uchida, I., *Anal. Chem.* **64**, 2642 (1992).
121. Suzuki, H.; Tamiya, E.; Karube, I., *Electroanalysis* **3**, 53 (1991).
122. Fiaccabrino, G.; Koudelka-Hep, M., *Electroanalysis* **10**, 217 (1998).
123. Feeny, R.; Kounaves, S. P., *Electroanalysis* **12**, 677 (2000).
124. Erickson, K.; Wilding, P., *Clin. Chem.* **39**, 283 (1993).
125. Lauks, I. R., *Acc. Chem. Res.* **31**, 31 (1998).
126. Vandaveer, W.; Fritsch, I., *Anal . Chem.* **74**, 3575 (2002).
127. Craston, D.; Jones, C.; Williams, D.; El Murr, N., *Talanta* **38**, 17 (1991).
128. Wring, S.; Hart, J., *Analyst* **117**, 1281 (1992).
129. Gyurcsanyi, R. E.; Rangisetty, N.; Clifton, S.; Pendley, B.; Lindner, E., *Talanta* **63**, 89 (2004).
130. Kovacs, G.; Peterson, K.; Albin, M., *Anal. Chem.* **68**, 407A (1996).
131. Figeys, D.; Pinto, D., *Anal. Chem.* **72**, 330A (2000).
132. Vilkner, T.; Janasek, D; Manz, A., *Anal. Chem.* **76**, 3373 (2004).
133. Hadd, A.; Raymond, D.; Halliwell, J.; Jacobson, S.; Ramsey, J. M., *Anal. Chem.* **69**, 3407 (1997).

134. Wang, J., *Talanta* **56**, 223 (2002).
135. Vandaveer, W. R.; Pasas Farmer, S.; Fischer, D.; Frankenfeld, S.; Lunte, S. M., *Electrophoresis* **25**, 3528 (2004).
136. Baldwin, R. P.; Roussel, T. J.; Crain, M.; Bathlagunda, V.; Jackson, D.; Gullapalli, J.; Conklin, J.; Pai, R.; Naber, J.; Walsh, K.; Keynton, R. S., *Anal. Chem.* **74**, 3690 (2002).
137. Liu, R. H.; Yang, J.; Lenigk, R.; Bonanno, J.; Grodzinski, P., *Anal. Chem.* **76**, 1824 (2004).
138. Albert, K. J.; Lewis, N. S.; Schauer, C. L.; Sotzing, G. A.; Stitzel, S. E.; Vaid, T. P.; Walt, D. R., *Chem. Rev.* **100**, 2595 (2000).
139. Chen, Q.; Wang, J.; Rayson, G. D.; Tian, B.; Lin, Y., *Anal. Chem.* **65**, 251 (1993).
140. Otto, M.; Thomas, J. D. R., *Anal. Chem.* **57**, 2647 (1985).
141. Beebe, K.; Verz, D.; Sandifer, J.; Kowalski, B., *Anal. Chem.* **60**, 66 (1988).
142. Forster, R. J.; Regan, F.; Diamond, D., *Anal. Chem.* **63**, 876 (1991).
143. Diamond, D.; Lu, J.; Chen, Q.; Wang, J., *Anal. Chim. Acta* **281**, 629 (1993).
144. Glass, R. S.; Perone, S. P.; Ciarlo, D. R., *Anal. Chem.* **62**, 1914 (1990).
145. Wang, J.; Rayson, G. D.; Lu, Z.; Wu, H., *Anal. Chem.* **62**, 1924 (1990).
146. Stetter, J.; Jurs, P. C.; Rose, S. L., *Anal. Chem.* **58**, 860 (1986).
147. Fielden, P.; McCreedy, T., *Anal. Chim. Acta* **273**, 111 (1993).
148. Pearce, T.; Gardner, J.; Freil, S.; Bartlett, P.; Blair, N., *Analyst* **118**, 371 (1993).
149. Newman, A., *Anal. Chem.* **63**, 586A (1991).
150. Freund, M. S.; Lewis, N. S., *Proc. Natl. Acad. Sci. USA* **92**, 2652 (1995).
151. Diamond, D., *Electroanalysis* **5**, 795 (1993).

INDEX

Analytical Electrochemistry, Third Edition, by Joseph Wang